SHIFTING BASELINES IN THE CHESAPEAKE BAY

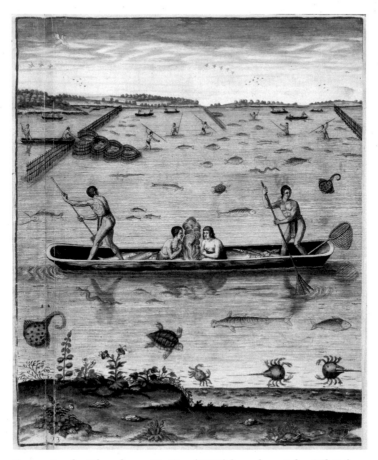

Engraving by Theodorus De Bry (1590) based on John White's watercolor of Indian fishing methods. Components of the illustration are discussed at www.virtualjamestown.org/images/white_debry_html/white.html#s46. Courtesy The Mariners' Museum and Park.

They haue likewise a notable way to catche fishe in their Riuers, for whear eas they lacke both yron, and steele, they fasten vnto their Reedes or longe Rodds, the hollowe tayle of a certaine fishe like to sea crabb insteede of a poynte, wherewith by nighte or day they stricke fishes, and take them opp into their boates. They also know how to vse the prickles, and pricks of other fishes. They also make weares, with settinge opp reedes or twigges in the water, which they soe plant one within another, that they growe still narrower, and narrower, as appeareth by this figure. Ther was neuer seene amonge vs soe cunninge a way to take fish withall, wherof sondrie sortes as they fownde in their Riuers vnlike vnto ours, which are alfo of a verye good taste. (Hariot 1590)

SHIFTING BASELINES

in the

CHESAPEAKE BAY

AN ENVIRONMENTAL HISTORY

Victor S. Kennedy

Johns Hopkins University Press • *Baltimore*

Johns Hopkins University Press
2715 North Charles Street
Baltimore, Maryland 21218-4363
www.press.jhu.edu

Library of Congress Cataloging-in-Publication Data

Names: Kennedy, Victor S., author.
Title: Shifting baselines in the Chesapeake Bay : An environmental history / Victor S. Kennedy.
Description: Baltimore : Johns Hopkins University Press, 2018. | Includes bibliographical references and index.
Identifiers: LCCN 2018002479| ISBN 9781421426549 (hardcover : alk. paper) | ISBN 1421426544
    (hardcover : alk. paper) | ISBN 9781421426556 (electronic) | ISBN 1421426552 (electronic)
Subjects: LCSH: Coastal ecosystem health—Chesapeake Bay (Md. and Va.) | Marine ecosystem
    health—Chesapeake Bay (Md. and Va.) | Chesapeake Bay (Md. and Va.)—Environmental conditions.
Classification: LCC QH541.5.C65 K46 2018 | DDC 577.5/1—dc23
LC record available at https://lccn.loc.gov/2018002479

A catalog record for this book is available from the British Library.

*Special discounts are available for bulk purchases of this book. For more information, please contact Special Sales
at 410-516-6936 or specialsales@press.jhu.edu.*

Johns Hopkins University Press uses environmentally friendly book materials, including recycled text paper
that is composed of at least 30 percent post-consumer waste, whenever possible.

Bill Black, Don Steele, Saul Saila, and Joe Mihursky
mentors during my early career
and
Deborah Coffin Kennedy
my companion by the Bay

Contents

Preface

In 2000, the Alfred P. Sloan Foundation supported a Census of Marine Life as a 10-year attempt to answer three Grand Questions:

What has lived in the oceans?
What now lives in the oceans?
What will live in the oceans?

The Census involved more than 2,700 scientists from 80-plus nations who used 14 field projects to explore these questions.[1]

To help focus on *What has lived in the oceans?*, the Census established the History of Marine Animal Populations program to apply concepts of historical ecology to fisheries management. The new program was "An interdisciplinary research program that used historical and environmental archives to analyze marine population data before and after human impacts on the ocean became significant." It involved historians, archaeologists, fisheries scientists, and marine ecologists in retrospective studies of formerly productive ecosystems and fisheries in European, Asian, Australian, and Pacific waters. Studies in the Americas focused on the Caribbean Sea, the Grand Banks of Newfoundland, and the Gulf of Maine Cod Fishery.

Surprisingly, the program did not examine the Chesapeake Bay, a particularly fertile ecosystem that supported some of the world's most productive fisheries in the late nineteenth century but that has since experienced a great decline. Throughout the later twentieth century, there was a general understanding that many of the Bay's fisheries were less bountiful than they had been, but the extent of those declines seemed not to be clearly appreciated as the century progressed. Although federal and state scientists had compiled numerous summaries of fishery data since the late 1800s, subsequent generations of fisheries managers and biologists seem to have generally overlooked or forgotten these earlier reports.

As an example of this generational amnesia, during the late nineteenth and early twentieth centuries, managers, resource scientists, watermen, and politicians in Maryland and Virginia had been concerned about Chesapeake Bay's oyster fishery and had attempted to reverse its decline (more on this in chapter 4). Later in the twentieth century, details about the historical extent of the fishery seemed vague. In 1978, scientists at the Virginia Institute of Marine Science produced an extensive report on the oyster fishery in Virginia that described historical abundances and changes over time.[2] To help understand the decline in Maryland so that it might be reversed, my colleague Linda Breisch and I in the early 1980s used numerous historical reports, both anecdotal and quantitative, to illustrate the past productivity of Maryland's oyster fishery and to document its decline.[3] Both these reports attracted great interest among managers, politicians, and biologists in the Bay region at the time.

This interest in the results of our historical sleuthing made me aware of how we can become fuzzy on details of what fisheries used to be like and how understanding those details can be important in supporting restoration. I subsequently expanded on management of the Bay's oyster fishery to include aspects of Virginia's fishery in an article in 1989,[4] then explored the history of fisheries beyond that of the oyster. In a book chapter on the history of the Bay's aquatic resources with Kent Mountford, I added information on commercial species in addition to oysters. In the mid-2000s, with Mike Oesterling and Willard Van Engel, I described the history of the US fishery for the blue crab, a species that still supported a vigorous fishery in the Chesapeake Bay when we wrote but whose numbers in the 1800s were expressed in superlatives. Finally, I have recently chronicled the US history of the diamond-backed terrapin fishery, once important but now closed in the Chesapeake Bay.[5]

As I was mining historical scientific papers, management reports, and newspaper articles to help write these stories, I was struck over and over by the remarkable abundances reported in the very early years of fisheries, in contrast to the diminished fisheries at the time when I was writing. Because the discipline of historical ecology can help us understand how the present depressed state of many aquatic resources came about and what that might mean for restoring these resources, I wrote this book to expand and consolidate my earlier work. The accounts of observers of the Bay's fishery resources from the seventeenth through the early twentieth centuries help us appreciate the past productivity of most of the Bay's dominant fisheries and understand how the fisheries have declined and why. Such knowledge can help boost strong efforts to restore fisheries.

Here is an example of how recognition of a collective amnesia about a fishery is leading to its renewal.[6] Biologists in the state of South Australia used historical records of catches by a fleet of oyster dredgers from about 1836 to 1910 as well as old reports by Inspectors of Oyster Fisheries to uncover evidence that reefs of the native oyster *Ostrea angasi* once flourished along 1,500 kilometers of the state's coastline. The fishery succumbed to overfishing and habitat damage and ended in 1944.

Oyster reefs in the area are now functionally extinct, and the fishery was forgotten. The biologists declared: "We suggest that the 180 years that have lapsed since fishing for *O. angasi* began and a lack of data . . . has contributed to a collective amnesia regarding the species' past distribution, abundance, and reefs. The current lack of consideration of this species represents a striking contrast to the attention it was paid historically." In late 2014, the biologists informed government scientists, managers, and industry representatives about the forgotten fishery, its former harvests, and estimates of the number of native oysters that reefs once supported. Stimulated by this evidence, South Australia's government and industry are again seeking to culture oysters in once-productive coastal regions.

Unlike in South Australia, where the oyster fishery had died and been forgotten, the Chesapeake Bay fisheries I describe are not forgotten. Some are in intensive care, even on life support, but I believe they can be nursed back to better health. Knowing how they once flourished can help those working to restore them set appropriate baselines or goals. When we understand the way we were and what we have lost, we can better support rehabilitation efforts.

Acknowledgments

My interest in historical ecology began when Dr. Rita Colwell, director of Maryland Sea Grant, asked my colleague Linda Breisch and me to provide a review of oyster biology. Convinced that you cannot manage what you do not understand, I wondered why Maryland's oyster fishery had declined and to what extent. Thus, I spent much time in the late 1970s in the McKeldin Library of the University of Maryland and the Eisenhower Library of the Johns Hopkins University examining early fishery reports and microfilmed newspaper articles and editorials. I appreciate Dr. Colwell giving Linda and me the chance to review eastern oyster biology and the history of its fishery.

The story of how extensive the Bay's oyster fishery once was and why it had declined so much stimulated my interest in the history of Bay fisheries. Fortunately, in recent years the scanning activities of Google and numerous library and educational institutions have provided electronic access to a multitude of online books and reports. This trove of historical material has saved me from commuting to College Park and Baltimore or from renting nearby apartments so that I could spend weeks or months browsing bookshelves and microfiches. I am very grateful for the existence of scanned archival material.

This work was prepared while I was a faculty member of Horn Point Laboratory of the University of Maryland Center for Environmental Science (UMCES). I am grateful to Dr. Mike Roman for the facilities made available to me. The final preparation of the manuscript was undertaken at Chesapeake Biological Laboratory, a facility that had played a key role in the early study of the Chesapeake Bay and its organisms. I am grateful to Dr. Tom Miller for office space and computer access.

Support for computer time used to track down references and prepare some original figures came from discretionary funds provided by the Keith Campbell Foundation for the Environment to UMCES President Dr. Donald Boesch, whose own moral support is much appreciated. Librarians Kathy Heil, Susie Hines, and Helen Cummings assisted in obtaining interlibrary loans. Robert Hurry of the Calvert Marine Museum pointed me toward a number of useful figures to illustrate the text. Deborah Coffin Kennedy drew some figures, and Anne Gauzens illustrated or modified others. Personnel in various agencies and historical societies gave permission to use images.

Chapters in draft were read critically by Deborah Kennedy and Caroline Wiernicki and by Drs. Walter Boynton, Lou Codispoti, Victoria Coles, Judy O'Neil, Jamie Pierson, Dave Secor, and Ryan Woodland. Two external reviewers provided additional helpful comments and suggestions. Tiffany Gasbarrini and Lauren Straley of Johns Hopkins University Press shepherded the manuscript through the publication process. Martha Sewall and her colleagues at the Press designed the text and prepared the book's cover. Copy editor Kira Hamilton ensured that issues of style and format were attended to while uncovering errors and inconsistencies in the narratives and citations. I am grateful for everyone's help. I take full responsibility for errors and misconceptions that may remain.

Units and Terms Used in the Text

Various measurement units have been used over the years in Chesapeake Bay fisheries. The English and approximate metric equivalents are given below.

UNITS OF MASS
Pound	=	454 grams, or 0.45 kilograms
Ton	=	2,000 pounds
Ton, metric	=	2,205 pounds

UNITS OF LENGTH
1 inch	=	2.54 cm
1 foot	=	30.5 cm, or 0.3 meters
1 yard	=	0.91 meters
1 mile	=	1.61 kilometers

UNITS OF AREA
1 acre	=	0.4 hectares

MEASURING OYSTER HARVESTS IN BUSHELS (KEMP ET AL. 1918, 74–80)
Maryland oyster bushel	=	2,801 cubic inches
Virginia oyster bushel	=	3,002 cubic inches
US standard bushel	=	2,150 cubic inches

Terms: Fishers (>90% men) in the Chesapeake Bay are called watermen (Warner 1976). Those who fish for crabs are called crabbers, categorized by their gear, so there are trotliners, potters, scrapers, and dredgers. A crabber may switch among gear and, when crab season ends, may become an oysterman and be a tonger or a dredger.

A Note on Anecdotal and Quantitative Harvest Statistics

Some historical numbers in this book may seem astonishingly large. Are they dependable or hyperbolic? Aside from a limited attempt by census officials to measure the country's fisheries in 1840,[1] there was no regular effort until the late 1800s by federal or state governments to either estimate natural abundances or assemble harvest statistics for commercial fisheries in the Chesapeake Bay. Thus, many accounts of harvests in years before the late 1800s were anecdotal, and anecdotes can be exaggerated. As an example, in 1835 Joseph Martin reported that 22,500,000 shad and 750,000,000 river herring were captured over six weeks by about 150 beach seine fisheries strung along the Potomac River's shores (see chapter 3). Some people in later years questioned the accuracy of the numbers. Clearly, the round numbers were estimates. Nevertheless, in 1889, US Fish Commissioner Spencer Fullerton Baird wrote that even if Martin's numbers were twice the actual numbers of fish harvested, they were still evidence of the formerly great abundances of shad and river herring in the Potomac.

Fortunately, some quantitative data exist for one Potomac fishery in the decades before Martin's report. In chapter 3, I present fishery scientist William Massman's review of the written daily records of fish harvested from 1814 to 1824 by George Chapman's fishery. The highest annual harvests were impressive—the 1814 fishery captured 180,755 shad, and the 1816 fishery landed 1,068,932 river herring. This was just one of many, many seine fisheries in the Potomac (Martin reported about 150 in 1835). All in all, the numerous anecdotal accounts of animal abundances that I report throughout this book tend to support each other, with the more quantitative accounts in later years supplementing the picture of formerly great abundances before fisheries began to decline.

To enable the federal government to study declining fisheries and recommend steps to restore them, the US Fish Commission was established in 1871.[2] In collaboration with the Tenth Census in 1880, the commission undertook the largest compilation of data on US fisheries to that date. Under the leadership of George Brown Goode, knowledgeable individuals, including about 30 trained statistical reporters, worked to compile data on the fisheries. The result was the publication between 1884 and 1887 of the monumental "The Fisheries and Fishery Industries of the United States"—10 volumes comprising 3,609 pages, 532 plates, and 49 charts of fishing grounds.[3]

Eventually the commission employed a permanent force of field agents trained to collect and compile statistics on all US fisheries. Justifying this focus, US Fish Commissioner Hugh Smith wrote in 1894: "The importance of statistics in general needs no demonstration, and the value of statistical information regarding the fishing industry is certainly as great as that of any other branch of human enterprise."

In his 1895b report on the fisheries in the Mid-Atlantic states, a source of many of my quantitative data, Smith stated that seven agents had accumulated the data on which he reported and that

> Whenever available, records have been consulted in ascertaining the quantity and value of the catch and in the case of a very large proportion of the professional fishing the figures presented may be regarded as being as nearly correct as it is possible to obtain. On the other hand, in the case of the semiprofessional fishing, especially that carried on in the upper courses of the rivers, it is the exception to find fishermen who keep a record of their catch, and in order to determine the approximate output of the various kinds of fishes taken it is often necessary to follow up very slight clues. A certain proportion of the fishermen know how much their fish sold for, and with this item as a basis the agents can, by judicious questioning, prepare a fairly accurate statement of the quantity of the yield.

My approach in this book is to bring together early anecdotes of the Bay's abundances, then turn to the reports of the US Fish Commission as well as those of Virginia and Maryland management agencies for later, more quantitative, accounts. The result is a picture of an ecosystem once teeming with fish, oysters, terrapins, wildfowl, and blue crabs in abundances that we can now only marvel at. We will see that the baselines by which we judge abundances have shifted downward over time.

SHIFTING BASELINES IN THE CHESAPEAKE BAY

Shifting Baselines in the Chesapeake Bay, the Immense Protein Factory

> Baltimore lay very near the immense protein factory of Chesapeake Bay, and out of the bay it ate divinely.
>
> Mencken (1940)

IN THE 1800S, the Chesapeake Bay was celebrated for its aquatic bounty. Within the Bay region, Mackay Laffan reported in 1877 that a "plain winter dinner in Maryland" included four small Lynnhaven oysters (from Virginia's Chesapeake Bay), terrapin à la Maryland, canvasback ducks, a small salad of crab and lettuce, baked Irish potatoes, fried hominy cakes, and plain celery (since this was a winter dinner, shad, a common spring delicacy, was missing from the menu).

Additional testimony to the region's epicurean reputation was given by Boston native Oliver Wendell Holmes in an 1860 essay. Asked by a young man what Holmes thought about Baltimore, Holmes took a jab at a Baltimorean's ability to think hard but excused it as follows:

> You are the gastronomic metropolis of the Union. Why don't you put a canvas-back duck on the top of the Washington column? Why don't you get that lady off from Battle Monument and plant a terrapin in her place? Why will you ask for other glories when you have soft crabs? No, Sir,—you live too well to think as hard as we do in Boston. Logic comes to us with the salt-fish of Cape Ann; rhetoric is born of the beans of Beverly; but you—if you open your mouths to speak, Nature stops them with a fat oyster, or offers a slice off the breast of your divine bird, and silences all your aspirations.

In the nineteenth century, shad and their river herring cousins, terrapins, sturgeons, and blue crabs, joined oysters and canvasback ducks in "stopping the mouths" of multitudes and supporting great industries. An extensive infrastructure of

Table 1.1 Persons employed in Chesapeake seafood industries over time

Year	Maryland	Virginia	Total	Reference
1880	26,008	18,864	44,872	Goode 1883b
1888	31,951	12,615	44,566	Collins 1891
1890	40,452	22,769	63,221	Smith 1895b
1891	39,944	23,595	63,539	Smith 1895b
1897	42,812	28,277	71,089	Townsend 1901
1901	36,260	29,325	66,685	Bowers 1907
1904	30,337	28,868	59,205	Bowers 1907
2014	8,332	14,618	22,950	NMFS 2016

boats, harvest gear, and processing facilities developed to exploit the Bay's abundant aquatic resources. Fishers harvested shad and river herring, oysters, terrapins, sturgeon, other fish, and blue crabs. Hunters shot ducks, geese, and swans for sale in cities. Processors salted and smoked shad and river herring, shucked and canned oysters, turned terrapins into gourmet food, plucked and cleaned wildfowl, prepared sturgeon steaks and caviar, picked crab meat, and nurtured soft crabs. Coopers made thousands of wooden barrels to hold salt fish. Tinsmiths handcrafted thousands of cans for oysters until machines took over to make tens of thousands in the same amount of time. Printers prepared labels for the cans, and women and girls pasted the labels in place. Net makers wove miles of netting for seines, gill nets, and pound nets. Carpenters built elegant sailboats for oyster dredgers and smaller sailboats for oyster tongers or crab trotliners. Sail makers supplied the sails and rope makers the ships' cordage and crab trotlines, while blacksmiths forged oyster dredges, crab scrapes, and oyster tongs. Railroads and steamships expanded their services into tidewater communities to collect raw and processed seafood for delivery to cities like New York, Montreal, Chicago, and San Francisco, as well as to ports for transport overseas to Europe, the West Indies, South America, and Australia.

The richness of the nineteenth-century fisheries supported work for thousands more people than now (table 1.1). As markets expanded, employment increased by about 58% between 1880 and 1897, a period when most Bay fisheries were near or at their peaks. Sadly, as the Bay's fisheries were overharvested, the bounty that had accumulated by natural processes over 10,000 years was squandered by humans in less than 100 years, diminishing employment.

The Bay's Productivity before and Just after Europeans Arrived

Scientists measure an ecosystem's biological productivity by estimating the amount of organic matter or energy content available during a given time period. As the last Ice Age waned, the Chesapeake Bay's productivity—the immense protein factory—developed and expanded, enabling Algonquian Indians living along tidewater to eat well.[1] In late autumn and winter, they hunted deer, bear, turkeys, and waterfowl. In March and April they used fishing weirs to trap anadromous fish that entered the Bay in spring.[2] Fish and oysters were smoked or dried for later consumption. By autumn, these preserved items, coupled with plants like corn, beans, and squash that were harvested in spring through autumn, were enjoyed during festivals celebrating the end of the nutritional year and anticipating the hunting that would resume shortly.[3] Tidewater Indian tribes are thought to have made limited inroads into the Bay's productivity, due mainly to their small population sizes (most tidewater settlements seem to have had fewer than 100 inhabitants).[4]

Like the Indians, early colonial settlers in the Chesapeake region farmed, hunted, and did some fishing, with their stores supplemented by helpful Indian groups. The settlers had sailed into a mature ecosystem rich with food resources.[5] Indeed, food might have been so abundant and so easily exploited as to make some diets monotonous. For example, while settler John Hammond extolled the positive aspects of Maryland and Virginia in 1656 in an essay written to counter negative accounts of the region, he admitted that "Deare all over the Country, and in many places so many, that venison is accounted a tiresom meat."[6] Similarly, foods like terrapins, wild ducks, and oysters were so common, so easily procured, and thus so inexpensive they were said to be fed to slaves and servants, perhaps more than the slaves and servants wished.[7]

This accessibility of what are now costly foods to poor people was more common when human population numbers were low and food was abundant and cheap. Charles Dickens in 1837 used the words of his characters Sam Weller and Sam's father to describe the apparent widespread availability of oysters and salmon to the poor in Britain in the 1830s:

'It's a wery remarkable circumstance, Sir,' said Sam, 'that poverty and oysters always seem to go together.' 'I don't understand you, Sam,' said Mr. Pickwick. 'What I mean, sir,' said Sam, 'is, that the poorer a place is, the greater call there seems to be for oysters. Look here, sir; here's a oyster-stall to every half-dozen houses. The street's lined vith 'em. Blessed if I don't think that ven a man's wery poor, he rushes out of his lodgings, and eats oysters in reg'lar desperation.' 'To be sure he does,' said Mr. Weller, senior; 'and it's just the same vith pickled salmon!'

Another illustration involves the common lobster from the North American Atlantic coast. Moses Henry Perley wrote in 1850 about a region of the Maritime provinces in Canada:

Lobsters are found everywhere on the coast, and in the Bay of Chaleur, in such extraordinary numbers, that they are used by thousands to manure the land. At Shippagan and Caraquett, carts are sometimes driven down to the beaches at low water, and readily filled with Lobsters left in the shallow pools by the recession of the tide. Every potato field near the places mentioned, is strewn with Lobster shells, each potato hill being furnished with two, and perhaps three, Lobsters.

In Britain today, oysters and salmon tend to be premium foods, generally beyond the daily reach of the poor. And while lobsters in the Maritimes and Down East are still abundant enough to be available as a special splurge for even a poor family, they are not so plentiful as to be taken from tide pools for use as fertilizer in potato fields. As human populations grow and wealth increases, overharvest of resources reduces their former abundances, making the resources scarcer and costlier. Such a decline has happened in the Chesapeake Bay, depleting the immense protein factory.

This book will illustrate the rise and fall of the Bay's fisheries, mainly in the nineteenth century. Understanding the decline involves understanding the historical ecology of the Bay, both to appreciate what has been lost in the fisheries and to help us think about what baselines (reference points) to use in restoring the system.

Because governments were slow to measure fishery harvests quantitatively, we must depend on anecdotes until the late 1800s. Nevertheless, the diverse sources and general agreement of these anecdotes support the idea of the former wealth of animals in the Bay. As the nineteenth century ended, it became clear to people living around the Bay that these astonishing abundances were declining. Thus, although US Fish Commissioner Smith was able to report in 1895 that "the great prominence which the fishing industry of these [Mid-Atlantic] states has attained may be said to depend on two products, namely the shad and the oyster, which are here more abundant and valuable than in all the remainder of the country combined," shad and oyster harvests continued to nosedive, as did harvests of the terrapin, the canvasback duck and other waterfowl, and the Atlantic sturgeon. These declines caused great contemporary concern, as is illustrated by historian Jennings Cropper Wise writing in 1911, just 16 years after Smith's report:

> Judging from the various statute books and court records of the seventeenth century, slight effort was made to protect the fish, oysters, terrapin and wild-fowl, all of which abounded in the waters of the Chesapeake and Atlantic Ocean on the Eastern Shore. So lavishly had nature stocked these waters with her delicacies that the supply was regarded as unlimited, and as usual no thought of the future was entertained until irreparable ravages began to show their effects. Thus is the improvidence of man wont to run its course and nature's well-nigh boundless stores are all but exhausted before human extravagance receives a check.

Here are some examples of the exhaustion of the Bay's "boundless stores." In 1896, more than 9,000 people, about 4,000 boats, and 19,000 nets supported the shad and river herring fishery (see chapter 3). Then, the annual shad harvest was about 17 million pounds; by 1980 when it closed in Maryland it was about 1 million pounds. The river herring harvest was about 30 million pounds in 1896 and about 1 million pounds in 1980. In the late 1880s, an estimated 15 to 20 million bushels of oysters were harvested in some years; today the harvest is a few hundred thousand bushels (see chapter 4). In the early 1800s, blue crabs were so abundant that there was no market for them; when they became entangled in the meshes of seine nets being used to capture fish, the seiners shook them out of the nets, crushed them, and left them to die on the shore (see chapter 8).

Baselines Begin to Shift

Can we restore the Bay to its former more productive state? People charged with restoring ecosystems use the concept of "shifting baselines" to describe a dilemma: What baseline should be used to compare past and present conditions of a system and assess how it has changed or might be restored?[8] A baseline tells us how things used to be, but ecosystems can deteriorate so gradually that each generation affected by the changes takes its present situation as the norm. Thus the baseline for the present generation is different from the baseline for the preceding generation, and so on back in time. The baseline—our perception—has shifted. To get a sense of what the Bay was like when fisheries were expanding to their peak, we must go back to the 1800s rather than just back as far as our own memory, or that of our parents or grandparents, takes us.

Two terrestrial examples of shifting baselines involve herds of bison "blackening the plains" and miles-long flocks of passenger pigeons "darkening the skies," but the reality of those plains or those skies is hard for us to imagine.[9] Our perception has shifted from that of our great-great-great grandparents who saw the herds or flocks. A similar but less commonly appreciated baseline shift has occurred in the Chesapeake Bay, as I will illustrate. Here are some seventeenth- and eighteenth-century reports of fish abundances to get us started.

In the 1624 account of Captain John Smith exploring the Chesapeake, the narrator described the richness of fish in the Patawomek (Potomac) region:

> in divers places that aboundance of fish, lying so thicke with their heads above the water, as for want of nets (our barge driving amongst them) we attempted to catch them with a frying pan: but we found it a bad instrument to catch fish with: neither better fish, more plenty, nor more variety for smal fish, had any of us ever seene in any place so swimming in the water, but they are not to be caught with frying pans.

The narrator then wrote that fish could be impaled on a sword, thereby improving on the use of frying pans:

> we spyed many fishes lurking in the reedes: our Captaine [Smith] sporting himselfe by nayling them to the grownd with his sword, set us all a fishing in that manner: thus we tooke more in owne houre then we could eate in a day.

Also in 1624, Reverend Alexander Whitaker described the seasonal arrival of enormous populations of anadromous fish species that moved into and through the Bay to spawn in upstream tributaries, responding to natural timing cues like temperature that still operate today; again, the fishing gear was not up to the task:

> The Rivers abound with Fish both small and great: the sea Fish come into our Rivers in March and continue untill the end of September: great sculles of Herings come in first: shads of a great bignesse, and the Rock-fish follow them. Trouts, Base

[Bass], Flounders, and other daintie fish come in before the others be gone: then come multitudes of great sturgeons, whereof we catch many, and should do more; but that we want good nets answerable to the breadth and deapth of our Rivers: besides our channels are so foule in the bottom with great logs and trees, that we often break our nets upon them.[10]

Father Andrew White mentioned shad, later to support a great fishery in the Bay, in his 1633 report: "One of these, named the Chesa-peack, is . . . navigable for large ships, and is interspersed with various large islands suitable for grazing; and at these islands can be caught, in the greatest abundance, the fish called shad."[11]

Boundless numbers of oysters and waterfowl also impressed early writers. One wrote in the 1600s about his ship almost running aground on oyster reefs that reached the Bay's surface (see chapter 4). Other accounts describe waterfowl covering the Bay's surface in huge rafts (see chapter 6). The wealth of fish, oysters, and crabs that was still available in the 1700s astounded Reverend Andrew Burnaby, who in 1775 marveled especially at the extraordinary numbers of sturgeon and shad in the Potomac River:

These waters are stored with incredible quantities of fish, such as sheepsheads, rock-fish, drums, white perch, herrings, oysters, crabs, and several other sorts. Sturgeon and shad are in such prodigious numbers, that one day, within the space of two miles only, some gentlemen in canoes, caught above 600 of the former with hooks, which they let down to the bottom, and drew up at a venture when they perceived them to rub against a fish; and of the latter above 5,000 have been caught at one single haul of the seine."[12]

It is important to understand that the Europeans celebrating the rich natural resources of the new world that they were exploring had sailed from a continent whose own resources had been enormously depleted and the environment degraded. As Callum Roberts wrote in 2007:

The rivers they sailed from in Europe were by this time awash with human waste, choked with sediment, and, in their upper reaches, blocked by long chains of milldams and weirs. Not since the early Middle Ages had Europe's major rivers run cool and clear. By the late fifteenth and sixteenth centuries, the days when shimmering columns of fish fought their way upstream to spawn were long forgotten.

Thus the anecdotes in this book reflect the appreciation of people whose own European baselines had shifted greatly and diminished over the centuries before the Americas became open for European business. What colonists saw was much different from what they had left behind, leading to enthusiastic superlatives in their descriptions. The following cartoon (figure 1.1) of a Chesapeake Bay oyster reef illustrates how baselines shifted from the days of early European colonization and through the nineteenth and twentieth centuries.

Figure 1.1. Cartoon of Bay oyster habitat in the 1600s before harvests became intense *(top)*, in the late 1800s when fishing intensities were very high *(middle)*, and in the early twentieth century when many fisheries had been greatly diminished *(bottom)*. Drawing by Anne Gauzens.

Developing Markets and Technology Helped Deplete the Bay's Resources

At least three factors contributed to the depletion of the Bay's abundant resources during the nineteenth century. One, described above, was the growth of human populations coupled with an increasing demand for seafood. A second was the push to develop markets, including seafood markets, beyond the Mid-Atlantic region. A third involved the invention of canning, with Baltimore becoming a major canning center.

With regard to the second resource-depleting influence—expanding markets beyond the Bay—transportation systems (roads, canals, railroads, steamships) grew to link coastal farming and fishing settlements to inland communities. There were many efforts to develop such transportation systems in the Chesapeake region in the early 1800s. Roads and canals came first. In 1808, Alfred Gallatin, President Jefferson's secretary of the treasury, produced a farsighted report that recommended a federally financed network of roads and canals to enable farmers to move their goods to market. Already, Baltimore had been connected westward to Cumberland, Maryland, by what was called the Bank Road (financed by banks) as a result of the state approving the linking of a number of private toll roads and turnpikes in 1805, with the network completed a few years later. Based on Gallatin's report, in 1811 the federal government began building the National, or Cumberland, Road by extending the Bank Road from Cumberland west to Wheeling, West Virginia. The road was completed in 1818, with an extension to Columbus, Ohio by 1833.[13] This network of roads (of crushed stone and gravel) facilitated the western transport of goods by vehicles such as horse-drawn Conestoga wagons.

Though obviously useful, the roads were expensive to maintain (heavy freight wagons tore them up, especially under wet conditions), and moving commerce over them was costly; roads were best used for short-distance deliveries.[14] Some Baltimore businessmen began considering alternatives such as canals, which could move materials 30 times more cheaply and cover longer distances than roads at the time.[15] However, Baltimore was hampered in competing with New York and Washington because it lacked a "decent river" (the Patapsco River beside which the city was built is only 39 miles long).[16] By contrast, the 315-mile-long Hudson River flowed from well inland to the New York City region, and Washington was connected by barge traffic on the 400-mile-long Potomac River to distant inland regions.

In the absence of a major river, a canal initially seemed a reasonable option for Baltimore because, by 1825, New York's Erie Canal had linked Albany on the Hudson with Buffalo on Lake Erie, sparking great enthusiasm for canal building nationwide. In that same year, construction had begun on the Chesapeake and Ohio Canal to link Washington to Cumberland.[17] Because Baltimore lies about halfway between the Susquehanna and the Potomac Rivers, two surveys in 1823 had explored the feasibility of connecting the city with either river by a canal, but both concluded that such canals would be costly and difficult to build.[18]

How then was Baltimore to ship the products of its emerging manufacturing and seafood industries westward? The answer to this question—a railroad!—was revolutionary, especially in light of the national enthusiasm about canals.[19] At that time the few railroads in the United States were used mainly for short hauls of bricks, gunpowder, or quarried stone by mules, horses, or gravity. Boldly, in 1827, 25 Baltimore businessmen resolved to build a railroad from Baltimore to Wheeling.[20] Their Baltimore and Ohio Railroad, the first US railroad to compete vigorously with canal systems, reached Cumberland in 1842 and Wheeling in 1853, enabling goods to move rapidly and relatively cheaply into the interior.[21] Thus, yearly shipments of oysters westward from Baltimore increased from about a half million pounds in 1848 to over three million pounds in 1860.[22]

The third factor helping to deplete the Chesapeake Bay's aquatic resources was the introduction of canning in the early nineteenth century. This technology began with the late-eighteenth-century application in Europe of heat sterilization for food preservation instead of depending on ice, salt, or smoke. In the United States, Thomas Kensett, Sr., is credited with the first use of cans for salmon, lobsters, and oysters in 1819, building a successful canning facility in New York.[23] In 1845, Edward Wright began canning oysters in Baltimore and continued for a few years before he died.[24] In 1849, Thomas Kensett, Jr., recognized that the bountiful Chesapeake Bay region was an important source of foods to be canned, so he moved to Baltimore, followed in time by other New England canners. Maryland farmers were cultivating peaches, apples, plums, and other fruits as well as an abundance of tomatoes, all perishable if not preserved. The Bay's wealth of oysters, crabs, and fish also required preservation for export. The new canning technology allowed Baltimore to become a center for preserving meats, fruits, and vegetables that were much in demand, not only on the East Coast and in western states and territories but also worldwide.[25] The technology was extremely efficient; oysters could be steamed, shucked, washed in ice water, packed in cans, steamed again, cooled, sealed, labeled, and packed in boxes ready for shipment in about an hour from the time they were offloaded at the dock for delivery to the nearby canning house.[26]

Canning became a profitable year-round business. Thousands of tons of fruit and vegetables were canned through the growing season from early spring until mid-September, and then oysters were canned from September to the next spring (figure 1.2).[27] To be close to the Bay's oyster resources, New England seafood processors opened branch plants in Baltimore and Norfolk in the mid-1830s (see chapter 4). As demand for Chesapeake oysters rose, the number of processing firms in Baltimore increased from one in 1834 to 80 in 1868.[28] Thus, by 1875 Edward King could report that the oyster trade in Baltimore was "stupendous," with whole streets devoted to oyster processing. Unfortunately, demand for the Bay's oysters eventually outstripped the supply (see chapter 4).

While the first attempt by census officials to measure the country's fisheries occurred in 1840, statistics were incomplete.[29] Nevertheless, they revealed that

Figure 1.2. Advertisement for a Baltimore firm that canned fruit and oysters in the late 1800s. Nichol 1937.

Maryland had at least 7,814 people working in fishing industries at the time (compare with later numbers in table 1.1), largely in oystering. Maryland also produced more than 70,000 barrels of pickled fish (probably shad and river herring), with Virginia adding over 20,000 more barrels. By 1860, Maryland, Virginia, New Jersey, and New York supported oyster industries that produced most of the million-plus dollars' worth of fishery products in the United States. The Civil War essentially closed Southern fisheries, including those in the Chesapeake Bay, but the fisheries rebounded rapidly after the war. However, the rebound was not accompanied by careful management. After an initial upsurge in harvests as the Bay's natural capital was being depleted, harvests declined sharply by the start of the 20th century.

Can We Develop Realistic Restoration Goals?

The immense protein factory needs to be restored, but what baseline do we use as a goal? To assist in choosing measures, I will visit historical accounts of the Bay's former abundances to help understand the way we were and to think about setting restoration goals. The focus is on shad and river herring, oysters, terrapins, wildfowl, and sturgeon, whose numbers are now a shadow of what they were in the nineteenth century. The blue crab is included because, although its harvests remain commercially productive today, its abundance in the 1800s was apparently much greater than now (see chapter 8), with possible effects of the crab's role in food webs (see chapter 9). Other accounts of the presently less depleted fisheries for commercial species, like menhaden and striped bass, are listed in the Further Reading section. Gear and methods used in the various fisheries are described and illustrated in an appendix.

Why the Chesapeake Bay Was So Productive and What's Changed

This noble bay and all its branches were alive with water-fowl and shellfish. Every point that jutted out into it was an oyster bar, where the most delicious bivalves known to the epicure might be had for the taking. Every cove, and every mat of seaweed in all the channels, abounded in crabs, which "shedding" five months in every year, yielded the delicate soft crab, and at every point on salt water, it was only necessary to dig along shore in order to bring forth as many mananosays, or soft shell clams, as one needed. It is a grave reflection upon the taste of our ancestors that there is no evidence earlier than the beginning of the present century [nineteenth] that the diamond-back terrapin was known or appreciated, but the more famous canvas-back duck certainly was known, and its qualities appreciated at a much earlier date.

Scharf (1879)

THE CHESAPEAKE BAY, the largest estuary in North America and one of the largest in the world, was and is renowned for its biological productivity. Its resource richness when European colonists arrived was described vividly by historian Thomas Scharf. Three centuries after the arrival of Europeans, scientists Samuel Hildebrand and William Schroeder were compiling a catalog of the Bay's fishes. They also described the Bay's productivity, comparing the fish production of the Bay in 1920 with that of Georges Bank, an oceanic habitat east of Cape Cod that supported the well-known cod, haddock, and other sea fisheries of New England: "It may be of interest to make a comparison here of the harvests of fish taken from Chesapeake Bay and Georges Bank, both intensively fished areas, the one protected by land and fed by numerous streams and the other in the open ocean. Chesapeake Bay and the brackish parts of its tributaries contain about 2,700 square miles and produced about 11 tons of fish per square mile in 1920, whereas Georges Bank, with

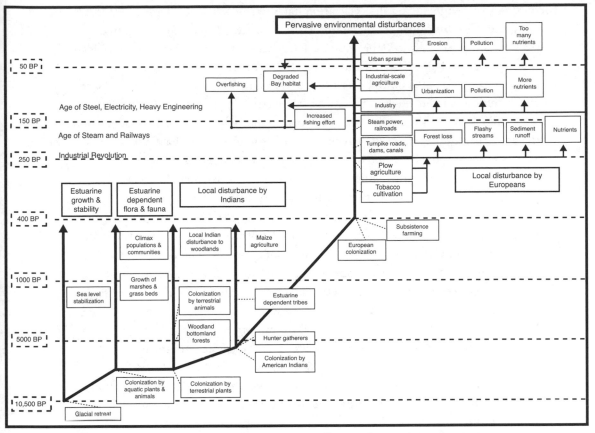

Figure 2.1. Simplified sequence of human events affecting Chesapeake Bay's aquatic resources. *Note:* Time intervals are not to scale. BP = years before present. Modified after Chesapeake Bay Fisheries Ecosystem Advisory Panel 2006.

an area of about 7,000 square miles, produced about 3 tons of fish to the square mile."

Of importance to my story is the fact that the Bay's fish production in 1920 that was celebrated by Hildebrand and Schroeder was a shadow of what it had been in the preceding century. The baseline had shifted to a lower level.

To understand the basis for the Bay's immense productivity, one must understand its geological, physical, chemical, and biological frameworks. And to recognize how we got to our present position of depleted aquatic resources, we have to understand the effects of human actions on these frameworks (figure 2.1).[1]

A Short History of the Bay and Its Interacting Ecosystems

The Chesapeake Bay is geologically young, having begun its life millennia ago as the Wisconsin Glaciation period ended and ice fields began melting (figure 2.1).[2] Ice formation worldwide during that glaciation had lowered sea level by over 300 feet, turning the offshore continental shelf of eastern North America into dry land. As the Wisconsin period ended, the massive Laurentide ice field, draped like an icy blanket over what is now Canada and the northern United States, began retreating northward. Its meltwater flowed over the coastal plain and down major East Coast rivers, including the Susquehanna and its tributaries, chewing into the shelf and carving valleys as it rushed to the sea at the continent's edge. Meltwater around the world gradually refilled the sea, which once again covered the East Coast shelf, slowly backing up into rivers and drowning river valleys about 10,000 years ago.

The water in the drowning valleys was salty where the sea met the river mouth, becoming increasingly fresh upriver as the sea's influence waned and the river's influence took over. The salt-fresh mixing regions formed estuaries. As estuaries grew and lengthened, they were colonized by aquatic plants and animals that tolerated brackish water conditions, and their shorelines were populated by terrestrial plants that provided food and shelter for land animals (figure 2.1). These newly emerging water-shoreline ecosystems were attractive to American Indians who used the high productivity to support their coastal populations.

Today's Bay is enormous, being nearly 200 miles long and with an average width of 12 miles, widening from north to south before narrowing toward its mouth.[3] The Bay and its tidal tributaries have nearly 12,000 miles of shoreline and a surface area covering about 4,500 square miles. Because of its size, the Bay's extensive drainage basin covers about 64,000 square miles and includes parts of six states and all of the District of Columbia (figure 2.2).

The Bay's geological structure includes a narrow central channel bordered by broad soft-sediment shoals. Although the channel reaches depths of 160 feet in some places, because of the shoals the Bay's mean depth is a relatively shallow 25 feet. The extensive shoals are important for the biological productivity of the system because the cycling or movement of nutrients between sediments and the overlying water is very efficient. In these shoals, sunlight is available to power the photosynthetic activities of algae and seagrasses that take advantage of the available nutrients. The shores are also fringed by extensive marshes that add to the ecosystem's plant productivity, which in turn supports animal productivity.

The Bay is affected by the three ecosystems that enfold it (figure 2.3), namely the overlying atmospheric ecosystem that delivers precipitation and material dissolved in rain and snow, the watershed ecosystem that distributes land-derived material in the inflowing freshwater tributaries, and the oceanic ecosystem that pumps marine material into and out of the Bay with each tidal cycle. Given the influences of these enfolding ecosystems, the Bay is a giant mixing bowl of fresh and salt water; of sediment particles; and of nutrients, fertilizers, and pollutants.

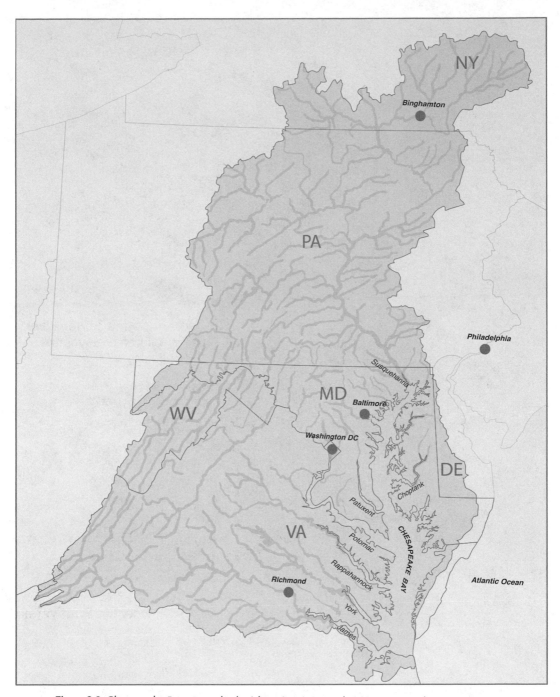

Figure 2.2. Chesapeake Bay watershed with major rivers and cities. Drawing by Anne Gauzens.

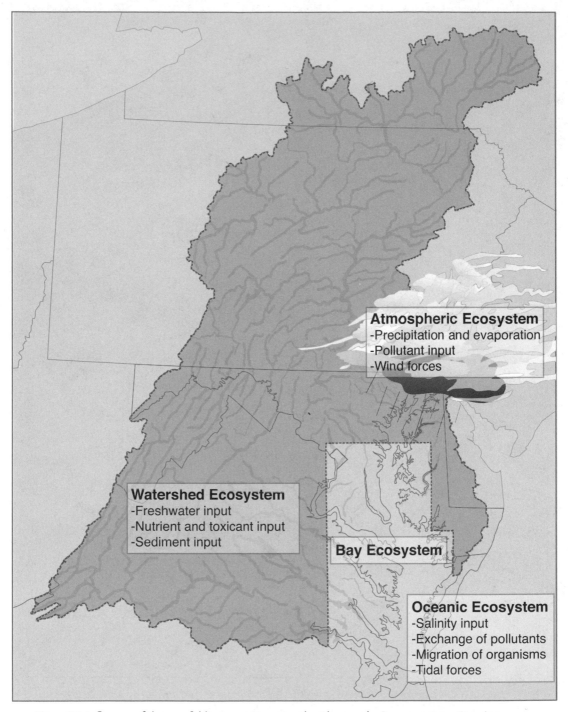

Figure 2.3. Influence of three enfolding ecosystems on the Chesapeake Bay ecosystem. Drawing by Anne Gauzens.

The Influence of Fresh and Salt Water

Rain or snow from the atmospheric ecosystem falls directly on the Bay itself and is also delivered by over 100,000 tributaries (rivers, streams, creeks) in the watershed ecosystem that feed into the Bay.[4] Most tributaries are relatively small, but three, the Susquehanna, Potomac, and James Rivers, provide about 80% of the Bay's freshwater input. Meanwhile, marine water from the adjacent ocean ecosystem keeps salinity high at the Bay mouth and for varying distances up the estuary, depending on a variety of factors described below. The result is a gradient in saltiness of the Bay water ranging from fresh water where rivers enter the estuary to oceanic salinities where the estuary meets the sea.

The gradient in salt affects different estuarine organisms differently, with some living most or all of their life in the estuary and others making seasonal use of the system. Anadromous fish species spend much of their adult life in the ocean but enter the estuary in spring and swim into low-salinity or fresh water to spawn, with their young returning to the sea to grow. Oysters can live in marine conditions and also tolerate relatively low salinities, but if salinity is too low they cease to feed and will not reproduce. Blue crabs spend most of their life cycle in the estuary; males and young can tolerate fresh water, but females with eggs move to the ocean, where the eggs hatch and the larvae live until stimulated to move back into the estuary as juveniles. Visiting waterfowl in winter can be found throughout the system from fresh water to the sea. Terrapins are limited to the estuary.

River inflow varies seasonally, with spring usually producing runoff from rainfall plus upstream snowmelt and with summer and autumn being drier. These variations affect the distribution of salinity within the estuary, with droughts allowing seawater from the ocean to push farther up into the Bay and high precipitation and snowmelt pushing back downriver and freshening the estuary. This back-and-forth between fresh water and marine flows has an important influence on the Bay's concentrations of sediments, nutrients, and pollutants. Over many parts of the Bay, it sets up a predominantly two-layer flow because fresh water flowing into the Bay from rivers, being lighter than sea water, floats on top of saltier water flowing into the Bay from the ocean (figure 2.4). The fresher and lighter surface water moves down-estuary, and the heavier and saltier bottom water moves up-estuary, with mixing occurring where the two layers meet. Buoyant particles are carried toward the sea in fresh water, but if they lose their buoyancy by clumping together or becoming coated with bacteria or algae, they may sink into bottom water before reaching the ocean and be carried back up the Bay and retained in the system. In this way, less-buoyant material may be kept within the estuary for some time rather than being flushed straight to the ocean.

Tidal ranges in the Chesapeake Bay are small, varying from about 2 feet at the head of the Bay to about 3 feet near the mouth. This limited range means that the rate at which water flushes from the Bay into the ocean is very slow, with fresher water remaining in the Bay for an average of seven months.[5] This slow flushing

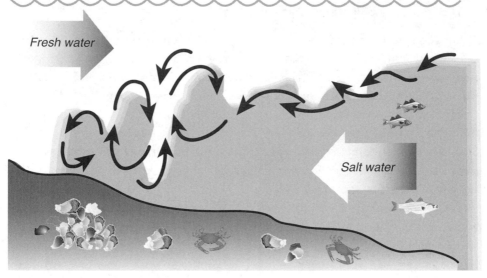

Figure 2.4. Mixing of lighter fresh water and denser salt water in an estuary.
Drawing by Anne Gauzens.

rate means that suspended and dissolved material may be retained in the estuary for a period of time, rather than being rapidly flushed out of the system, except under the most extreme rainfalls or snowmelts. However, retention of pollutants, excessive nutrients, and smothering sediments can be detrimental to an estuary and its associated organisms.[6]

The Influence of Sediments

Geologists refer to estuarine habitats like the Chesapeake Bay as depositional systems that receive and retain sediments brought in from the land (the watershed ecosystem) and from the sea (the oceanic ecosystem). The Bay collects mostly finer particles from terrestrial sediments eroded from the land by natural and human activities (figure 2.1) and carried by the Bay's tributaries as well as coarser sand pumped in from the ocean.[7] The sediments sink and cover the floor of the estuary. Hard surfaces are uncommon, yet some estuarine creatures like oysters need hard surfaces that they can attach to.

Before humans began providing hard surfaces like wharf pilings, riprap, sunken boats, and trash, hard surfaces were limited to any rocks protruding from the mud; to the shells of clams, mussels, and oysters; and to woody elements like trees and brush ripped from the land by storms and erosion (see Whitaker's complaint about submerged net-destroying logs and trees on page 6). Oyster beds thrive in the shallows where floating microscopic algae (phytoplankton) are readily available to be oyster food. Their shells provide hard surfaces for organisms to live on, including oyster

1. Oyster Spat
2. Skilletfish
3. Hooked Mussels
4. Whip Mud Worms
5. Sea Squirts
6. Sea Anemone
7. Barnacles
8. Fan Worms
9. Mud Crab

Figure 2.5. Oysters serving as habitat for a variety of organisms requiring hard surfaces to attach to and for others seeking shelter. From *Life in the Chesapeake Bay,* courtesy A. J. and Robert Lippson.

spat (settled larvae), and the spaces among shells provide shelter for other organisms (figure 2.5).

Trends in sediment deposition can be estimated by driving metal or plastic cylinders into the Bay bottom to collect sediment cores for analysis in the laboratory. Radiological and microscopic techniques estimate when sediments were deposited at different depths within a core. Such estimates reveal that sedimentation rates averaged about 0.2 inches per decade before Europeans colonized the region, rising to slightly more than 1 inch per decade when land clearing for farming was at its height in the late eighteenth and nineteenth centuries.[8] These rates imply that, while Indians had cleared some land for villages as well as for agriculture and to make hunting easier, their relatively small population and their nomadic habits meant that their influence on sediment production was slight. European settlers began clearing the forested watershed for the same reasons of habitation and agriculture. However, settlers did not move from dwelling site to dwelling site, so the land they cleared often did not have a chance to lie fallow and be revegetated. Sediment thus ran off harvested farm fields more rapidly than from vegetated land.

Early evidence of how transparent Bay water may have been before farming expanded is perhaps shown in De Bry's 1590 illustration of Native Americans fishing (see frontispiece). Down through the water we see a ray, a sturgeon, and some horseshoe crabs, creatures that are predominantly bottom dwellers, so we might assume that they could be seen through the water surface. Centuries later, James Hungerford described Bay water transparency in his 1859 novel about his experiences in 1832 as a young Baltimore lawyer visiting his uncle's plantation on Hungerford Creek, which flows into the lower Patuxent ("Clearwater"): "of all the bright rivers which flow into that noble bay [the Chesapeake] there is not one which excels the Clearwater in the purity of its waters . . . So transparent are its waters that far out from the shore you may see, in the openings of the sea-weed [seagrass] forest on its bottom the flashing sides of the finny tribes as they glide over the pearly sands."

Such water clarity is uncommon in today's Bay. As European populations increased, more and more land was disturbed by plowing or cleared for tobacco farming (figure 2.1). Today, about 58% of the sediment entering the Bay comes from farmland.[9]

The Influence of Nutrients

Nutrients (especially nitrogen and phosphorus) delivered by precipitation and river and ocean inflows contribute to high plant productivity—phytoplankton that float in the water, sea grasses attached to the bottom, and microalgae that live on sea grass leaves or the estuary bottom. Clear water allows light to penetrate deeply, powering photosynthesis in plants in the water and on the bottom (figure 2.6, top).

Excess amounts of nutrients support phytoplankton blooms in the surface water, limiting light penetration below the blooms, which hinders photosynthesis of deeper-dwelling plants whose decreased productivity in turn affects dependent animals.[10] By the mid-1980s the Bay was receiving about 8 times more nitrogen and 13 times more phosphorus than when colonists arrived, stimulating excessive algal growth.[11] One major effect of algal hyperproduction is an increased occurrence and extent of low-oxygen (hypoxia) or no-oxygen (anoxia) events in the Bay's deeper regions (figure 2.6, bottom).[12] These "dead-zone" conditions occur because the increased amounts of phytoplankton that are not eaten by animal grazers eventually die, sink to the estuary bottom, and decay. Their decay is caused by microbes, which respire, using up the available oxygen in the bottom water. Organisms that cannot move away from the affected regions, like oysters on reefs or creatures that live buried in the soft sediments, may die from lack of oxygen, thus disrupting food-web relationships.[13]

Oysters are filter feeders that eat phytoplankton, and past abundances of oysters helped control algal populations. However, the extreme decline in oyster populations to be described in chapter 4 minimized the filtering effect of this major consumer of phytoplankton.

The Influence of Pollutants

As described thus far, the Chesapeake Bay has experienced substantial human-induced changes since European colonists arrived over 400 years ago (figure 2.1).[14] Up to 18 million people now live in the watershed, a huge increase from the 2 to 4 million thought to have inhabited the region in the late 1880s when the active exploitation of Bay resources was at its peak. Twenty million people are expected by 2030.[15]

These millions affect the Bay directly or indirectly by their activities. Most people live in major metropolitan areas, but no matter where they live in the watershed, people often dispose of their domestic and industrial wastes into a drainage basin

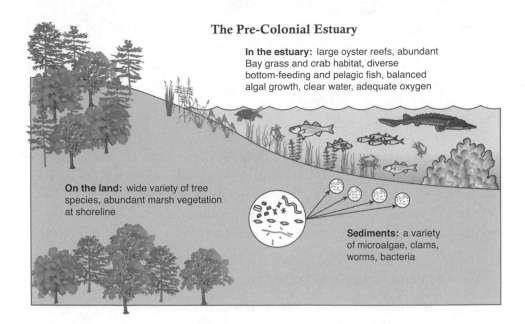

The Pre-Colonial Estuary

In the estuary: large oyster reefs, abundant Bay grass and crab habitat, diverse bottom-feeding and pelagic fish, balanced algal growth, clear water, adequate oxygen

On the land: wide variety of tree species, abundant marsh vegetation at shoreline

Sediments: a variety of microalgae, clams, worms, bacteria

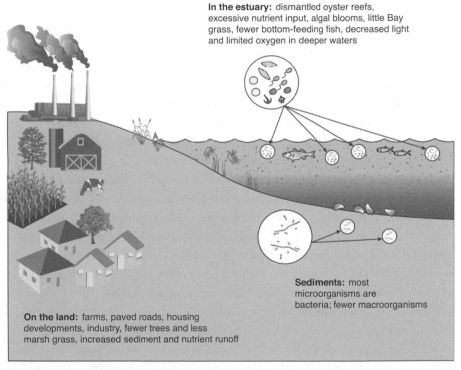

The Post-Colonial to Modern-day Estuary

In the estuary: dismantled oyster reefs, excessive nutrient input, algal blooms, little Bay grass, fewer bottom-feeding fish, decreased light and limited oxygen in deeper waters

Sediments: most microorganisms are bacteria; fewer macroorganisms

On the land: farms, paved roads, housing developments, industry, fewer trees and less marsh grass, increased sediment and nutrient runoff

Figure 2.6. Changes in Chesapeake Bay between pre-Colonial times (*top*), when land plants (forests, marshes) minimized sediment and nutrient runoff into the clear waters of the estuary, and post-Colonial times (*bottom*), when land had been cleared and "developed," resulting in increased sediment and nutrients draining into the increasingly murky estuary. Courtesy Maryland Sea Grant from Brush 2017, with modifications.

that ultimately feeds to the Bay. Wastes enter the Bay through three main sources. The first involves nonpoint sources such as agricultural runoff and discharge from septic systems, whose waste material diffuses through the soil to rivers and the Bay. The second includes point sources such as the outlets of sewage treatment plants or industrial facilities. The third source involves wastes delivered to the Bay while dissolved in rain or snow.

Wastes from these three sources may include pollutants.[16] Farmers use fertilizers, pesticides, and herbicides that if applied carelessly can run off into the Bay and damage aquatic plants and animals. Homeowners seeking the perfect lawn or garden use similar substances. The increased input of sediments and chemicals has concerned scientists, managers, environmentalists, and politicians since the 1970s, and a variety of agreements and directives to respond to these concerns has been implemented.[17]

The Additional Influence of Technological Revolutions

Clearly, sediments, nutrients, and pollutants may play important roles in harming the Bay's ecosystem, which has an effect on the ecosystem's creatures. However, this book focuses on the role of excessive exploitation of once-abundant animal resources. The increased demand for food as human populations grew not only involved a move to industrial-scale farming on land but also intensified fishing (figure 2.1). Fishing grew from subsistence to industrial levels as a number of technological revolutions occurred.

Historians and economists have identified five great technological revolutions, or "surges of development," that sparked dynamic changes in transportation, industry, and even fisheries (figure 2.1).[18] The three surges that affected early exploitation of the Bay's aquatic resources were the Industrial Revolution from the late 1700s, the Age of Steam and Railways of the early 1800s, and the Age of Steel, Electricity, and Heavy Engineering of the late 1800s (figure 2.1). The primitive roads of the late 1700s were gradually improved in the early 1800s and were supplemented by canals.[19] Thus the Cumberland Road and the Chesapeake and Ohio Canal enabled Bay seafood to be marketed inland (see chapter 1). As steamships and railroads developed during the Age of Steam and Railways, seafood markets expanded even farther. The cheap steel of the Age of Steel, Electricity and Heavy Engineering allowed speedy steamships to be built and worldwide shipping to grow, with Bay seafood reaching Europe and even Australia, stimulating greater exploitation of aquatic resources. The contemporaneous advent of electricity allowed canning industries to expand their seafood processing activities.

As these surges of development unfolded, land clearing and industrial growth coupled with overfishing caused the Bay's environment to deteriorate (figure 2.1). Fish spawning habitat was polluted or mined for sand or gravel. Oyster reefs shrank and were silted over. Fish and shellfish populations declined greatly, leading to shifting baselines in the memories of Bayside residents. Such declines in fishery productivity will be described in subsequent chapters.

The Spring Fishery for Shad and River Herring

A Hectic Scramble

~~~~~~~~~~

> In the spring of the year herrings come up in such abundance into their brooks and fords to spawn, that it is almost impossible to ride through without treading on them . . . Thence it is that at this time of the year the freshes of the river, like that of the Broadruck, stink of fish.
>
> Beverley (1722a)

IN THE SPRING OF the Chesapeake year, a number of shrubs and small trees called shadbush become covered in snowy-white flowers. Their name celebrates the coincidental appearance of adult shad and their river herring cousins (figure 3.1) in the region.[1]

These anadromous fish live in the coastal Atlantic Ocean during other seasons of the year, responding in spring to environmental cues like temperature and entering estuaries to swim upstream to the habitats where they will spawn.[2] After spawning, adults that have not died or been eaten by predators return to the ocean, to be followed late in the fall by their surviving offspring (figure 3.2).

Anadromous fish have probably been springtime food for humans since the latter arrived on the shores of the Bay and its tributaries. Indians along the banks of the Susquehanna ate large quantities of shad[3] that were caught in a number of ways, according to Meehan (1893): "weirs and traps; sieves [seines], gill and scoop nets; spears, bows and arrows and gigs; hand, pole and set lines. They even knew how to stupify fish by using intoxicating substances. Besides these things they constructed pens and preserves in which fish could be kept alive until wanted." Colonists also employed these devices to capture fish. Gill nets, seines, and pounds (see figures A.1–A.5) were the most important in the shad fishery.

Figure 3.1. Shad *(left),* alewife *(middle),* and blueback herring *(right).* National Museum of Natural History.

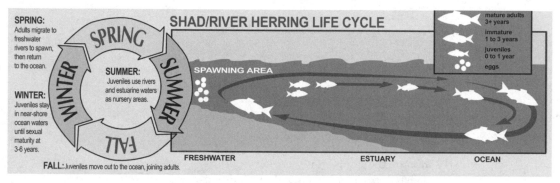

Figure 3.2. Life cycle of shad and river herring. Modified with permission from Pennsylvania Fish & Boat Commission.

The shad's seasonal appearance was echoed in some Indian names; the Lenape called March "the month of the shad," and the Algonquians named the moon that occurred during the peak April migration of shad "Spearfish Moon."[4] European writers in the seventeenth and eighteenth centuries mentioned the seasonality of these fish as well as their size and abundance. Three accounts in particular are instructive (see also quotes by Whitaker 1624 and Burnaby 1775 in chapter 1):

> For foure monethes of the yeere, February, March, Aprill and May, there are plentie of Sturgeons: And also in the same monethes of Herrings, some of the ordinary bignesse as ours in England, but the most part farre greater, of eighteene, twentie inches, and some two foote in length and better; both these kindes of fishe in those monethes are most plentifull, and in best season, which wee founde to bee most delicate and pleasant meate (Hariot 1590).
>
> Shaddes, great store, of a yard long, and for sweetnes and fatnes a reasonable good fish, he is only full of small bones, like our barbells in England (Strachey 1610).
>
> When they spawn, all streams and waters are completely filled with them, and one might believe, when he sees such terrible amounts of them, that there was as great a supply of herrings as there is water. In a word, it is unbelievable, indeed, indescribable, as also incomprehensible, what quantity is found there. One must behold [the sight] oneself (Byrd 1728).

Early Fisheries for Shad and River Herring

Like the Indians before them, European settlers took advantage of this abundant food supply. Famously, George Washington wrote to a friend that the Potomac shore of his Mount Vernon estate was "washed by more than ten miles of tidewater," that "several valuable fisheries appertained to it," and "the whole shore, in short, was one entire fishery." Over the years, Washington paid careful attention to mesh sizes of the seines he purchased, the quantity of fish harvested annually over the years, and how they were to be preserved or sold. He admonished his workers to put part of the early run of fish aside to ensure a backup supply for his slaves if the run proved a bust.[5]

Almost a century later, James Milner reported in 1876 on the seine fisheries of the Potomac and Patuxent Rivers, where the numbers of captured fish were lower than in previous years. The largest seine being used was at Stony Point, Virginia, on the Potomac. The seine with its lines was nearly 5 miles long, with the net itself being 9,600 feet long. An earlier long-running fishery (using a much smaller seine; see below) was located at Chapman's Point, Maryland, in the Potomac. Milner reported that Mr. Chapman, owner of the fishery, reminisced about a time past "when the seine-hauls on the shore piled the herring up from the water's edge 12 or 15 feet landward. The men walked or waded knee-deep among them, thrusting in their arms to find and select out the shad, and allowed the herring to float off at high tide." When he spoke in the late 1870s, Mr. Chapman called this practice a "reckless, destructive policy" that fishermen now regretted because of the scarcity of fish.

In 1961, fishery scientist William Massmann summarized the records kept from 1814 to 1824 by Chapman's fishery. Slaves hired from their owners hauled the 1,050-foot seine every day in season (late March to mid-May). Annual harvests ranged from 27,939 to 180,755 for shad and 343,341 to 1,068,932 for river herring. Massmann calculated that 955,615 shad were landed during in the 11-year period of this fishery (one of many in the river), an amount equal to about one-third of the shad harvests by all gear in the Potomac from 1946 to 1956, 132 years later. Based on Chapman's careful records, Massmann concluded that the quantities of shad harvested by the numerous Potomac fisheries in the early 1800s would be "prodigious" and that "These records also suggest that reports by early historians concerning the tremendous quantities of fishes found then may not have been exaggerated." One of those early historians was US Fish Commissioner Marshall McDonald, who in his 1887 review of the fisheries of the Chesapeake Bay wrote: "The Potomac has always been celebrated for the excellency and value of its shad and herring fisheries. Reports of their magnitude have come to us from early days, and from them we gather that the production must then, as compared with our own day, have been simply fabulous."

Early numbers of anadromous fish in tributaries other than the Potomac were also very high. Virginia Fish Commissioner Alexander Moseley wrote in 1877 about the fish present in the James River watershed of 12,000 square miles that was home to one-third of the Commonwealth's human population: "Within the memory of men still living, this large population could easily obtain one-half of its annual subsistence from the fish [shad and river herring in spring, striped bass in autumn], fresh or salted, supplied by the waters of the main stream and its various tributaries. It was a common saying in those days, that the shad and herring were equal to half a hog crop."

Farther up the Bay, the Susquehanna had also supported important fisheries in the past. There is anthropological and narrative evidence that Indians took great quantities of shad in the area before settlers arrived.[6] Settlers followed suit. Pennsylvania Fish Commissioner John Gay wrote in 1892 that there were about 40 fisheries along the banks of the North Branch of the Susquehanna between Northumberland and Towanda, Pennsylvania, in the 1800s. Employing the adjective "fabulous" once more but backing up its use by referring to "authentic data," Commissioner McDonald wrote in his 1887 review of Bay fisheries:

> The accounts in regard to abundance of shad, given by the early settlers on all the
> rivers of the Atlantic slope, seem almost fabulous to us in these days. If tradition
> has invented exaggerated stories concerning all the other rivers, those accounts
> touching the Susquehanna at least are undoubtedly established by authentic data.
> At the request of Prof. S. F. Baird, U.S. Commissioner of Fish and Fisheries, a
> committee of the Wyoming Historical and Geological Society, of which Mr. Harrison
> Wright was chairman, has prepared and submitted the following very interesting
> report on the early shad fisheries of the North Branch of the Susquehanna River.

Harrison Wright's comprehensive report that McDonald cited appeared in 1884 and was based on multiple interviews of old settlers and readings of county records, old newspaper files, and printed histories. It included a letter written in 1881 by Gilbert Fowler of Berwick, Pennsylvania, on his eighty-ninth birthday. Fowler described the fruitful fisheries of his youth, when 11 to 12 thousand shad could be caught in one seine haul. He provided this astonishing account of bow waves pushed up by the onrushing shad:

> The next fishery [on the Susquehanna] was that of Samuel Webb, located about four
> miles this side of Bloomsburg [Pennsylvania]. This was an immense shad fishery.
> From the banks of the river at this fishery could be seen great schools of shad
> coming up the river when they were a quarter of a mile distant. They came in such
> immense numbers and so compact as to cause or produce a wave or rising of the
> water in the middle of the river, extending from shore to shore. These schools,
> containing millions, commenced coming up the river about the first of April and
> continued during the months of April and May . . . The Susquehanna shad constituted the principal food for all the inhabitants. No farmer or man with a family was
> without his barrel or barrels of shad the whole year round.

In addition to the North Branch fisheries, numerous other seine fisheries existed in the Susquehanna.[7] In the early 1800s before dams impeded their progress, anadromous fish swam north to Binghamton, New York, a journey of over 500 miles from the Bay mouth.[8] The fish would arrive there about the end of April, and the fishery would persist through May. Gear was mainly seines measuring 265 to 495 feet long and managed by six to eight men; several hundred fish could be captured in a sweep.[9] Farther downstream in later years, seines might be up to two miles long. Because some Susquehanna tributaries lacked beaches suitable for seine hauling, or because many suitable beaches were already spoken for by other seiners, some entrepreneurs built shad floats—large floating structures moored offshore in the river, with sloping decks up which seines could be dragged.[10]

The vigor of the migrating fish that created bow waves in their surge upriver was matched by the vigor with which nineteenth-century humans in the Chesapeake watershed prepared for their capture. Coopers made barrels to hold the salted catch, merchants collected salt to preserve the fish, and fishers mended nets and prepared their boats, winches, and floats. For an intense six to eight weeks, thousands of people caught and salted, smoked, or pickled fish, with some people temporarily leaving their regular job to join the fishery. This fishery was at one time the second-most economically important in the Chesapeake region, after the oyster fishery.[11]

An illustration of the intensity of the fishery is given by Joseph Martin, who wrote in 1835 about the spring fishery in the Potomac River that "quantities of shad and herrings are taken, which may appear almost incredible"; 4,000-plus shad and up to 300,000 herring might be captured in a seine haul. He specifically noted that in 1832, a single seine haul landed "a few more than 950,000 [herring] accurately counted." (see A Note on Anecdotal and Quantitative Harvest Statistics). Martin drew up a table in 1835 for the port of Alexandria to show the economic effect of the appearance of shad and river herring in the Potomac River region in the early 1830s (table 3.1). River herring were sometimes so abundant that there was no market for them, and they were given away or used as fertilizer by farmers.

Over time the Susquehanna fishery declined, as demonstrated by the 1895 report of magazine writer David Fitzgerald that whereas a seine haul could trap 10,000 fish in previous decades, as years passed catches in a single haul gradually dropped to 5,000, then to 1,000, then to 500 when he wrote. Although the overall pattern was of declining harvests, fish populations might vary greatly from year to year for

Table 3.1 Shad and river herring fisheries on Potomac River shores in the early 1830s

Number of fisheries on the Potomac	~150
Number of vessels	450
Number of men on the vessels	1,350
Number of laborers on shore	6,500
Number of shad landed in six weeks	22,500,000
Number of herring landed	750,000,000
Bushels of salt for curing the catch	995,000
Number of barrels used	995,000

Source: Martin 1835.

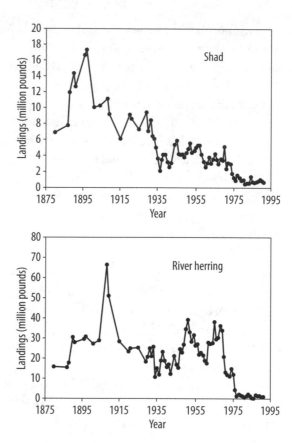

Figure 3.3. Landings of shad (*top*) and river herring (alewife and blueback herring, *bottom*) throughout the Chesapeake Bay. Data from Fisheries Statistics Division 1990.

unknown reasons. Naturalist Lewis Radcliffe wrote in 1922 that in years like 1867 and 1873, Potomac River shad harvests were much larger than harvests in between those years, with an unexplained large decline from 1873 to 1878. Twenty years after that period of decline, Pennsylvania Fish Commissioner William Meehan reported in 1898: "the herring fisheries in the Susquehanna this year were much larger than in former years . . . for the vastness of the schools in the extreme lower part was beyond compare. As it was, the nets in the Pennsylvania section took thousands upon thousands more than could be used."

Nevertheless, despite the occasional good years, the overall trend in harvests Bay-wide was downward. Shad harvests peaked in 1897, declining thereafter, and river herring harvests peaked in 1908 before waning (figure 3.3). Diminished harvests prompted searches for causes and remedies.

Factors Leading to Diminishing Populations

Schoolteacher and historian John Wilkinson stated in 1840 that shad had not reached Binghamton for 12 to 15 years, and 58 years later fisheries statistician Charles Stevenson could report that shad no longer reached beyond Clarks Ferry (about 234 river miles downstream from Binghamton). Stevenson blamed artificial dams that blocked fish migration upstream, river pollution by sawdust and chemicals, and extensive capture downstream of migrating fish for these changes in upstream distributions in the Susquehanna and elsewhere. For example, rivers in the Chesapeake watershed were dammed from the late 1600s to form ponds to "run iron forges, furnaces, mining operations, and . . . mills."[12] Furthermore, the desire in Pennsylvania to emulate the commercial success of New York's Erie Canal stimulated construction of canals and dams that fed the canal or formed slackwater pools to allow passage across shallow, rapid-rich rivers like the Susquehanna from one canal to another. Furthermore, pollution from sawdust and industrial chemicals from streamside industries smothered spawning beds or repelled fish.

The Susquehanna was not the only river whose migrating spawners once nourished humans living upstream, and the deleterious factors reported by Stevenson that hindered annual migrations in the Susquehanna affected other tributaries as well. Thus the Virginia Commissioners of Fisheries' *Annual Report of 1875* noted that scarcities of anadromous fish had long concerned Virginians living along tributaries in which there were hindrances to fish movement upriver: "It was not until 1740 that the whole colony became agitated about the scarcity of fish and the obstructions to their passage, and from that time till the war of the Revolution, the Legislature was crowded for petitions for the destruction of dams, hedges, stone stops, &c."

The commissioners also poked mild fun at excuses put forth by different groups of fishers to explain fish scarcity:

Various causes were assigned by the different and conflicting Interests. The shore fishermen ascribed it to the gill-nets, which they said invaded the spawning-beds [and] left no place unmolested where spawn was deposited. The gillers, on the other hand, protested that the injury was done by the draw-seines, which scraped the bottom, while they merely skimmed the surface. But both sets concurred in ascribing the greatest and most serious damage to the pound and staked nets and other murderous contrivances, which almost barricade the river near its mouth, and prevent the ascent of fish. "They should all be prohibited by law," is the joint cry. When we descended among the pound and staked nets, we heard a very different music. There was no need of any protection or prohibition—they were merely catching the offal, and letting all the good fish ascend the stream, &c. Diverse interests inspire diverse sentiments.

The commissioners decided that four factors were important in depleting fish numbers upstream and suggested solutions as follows:

(1) Dams across the rivers prevent fish from ascending to the best spawning grounds. It is an ascertained fact, that of the eggs deposited high up a stream, a much larger proportion hatches and survives than when deposited lower down, and

that the young fish attain greater vigor and strength by the time they reach tide, and are better fitted to encounter its fierce current.

(2) Excessive seining occurs below and above tide. There should be a close time each week of forty-two hours at least—from Saturday, noon, to Monday morning at sunrise—in all tide-water; and above tide, no seine or net whatever should be allowed. In the smaller streams in the interior, one day's seining will almost exterminate the race.

(3) Staked and pound nets, traps and other fixed and stationary contrivances lead to the wholesale destruction of fish and should be sternly interdicted in the Commonwealth.

(4) The refuse from gas works, paper mills, &c., is also very destructive to fish, or is very distasteful to them, that they abandon the river. Sawdust from sawmills is said to depopulate streams rapidly, and all should come under prohibition.

The commissioners concluded hopefully that the fishery in Virginia could be saved and improved: "We saw enough, however, to satisfy us that by artificial propagation [fertilizing eggs from captured females and rearing fry in hatcheries], clearing the stream of all obstructions to its source, and a few wholesome regulations as to close time, size of mesh and fixed nets, the river before us might be made to yield food for millions."

By the end of the nineteenth century, although biological factors governing population dynamics might have been incompletely understood, it was clear that dams, damaged spawning habitat, and the gauntlet of nets the fish had to run to reach their spawning grounds continued to affect the fishery. Thirty-nine years after the Virginia Commissioners of Fisheries wrote about these matters, US Fish Commissioner Hugh Smith in 1914 illustrated the changes over time (i.e., the shifts in baselines) in Bay fisheries for shad and river herring by reporting for a Potomac River seine fishery:

At Ferry Landing, Va., the largest seine on the river, 1,200 fathoms long [7,200 feet], discontinued operations in the middle of the season [1913] owing to the scarcity of fish. In former years this celebrated fishing shore, with a smaller seine, sometimes yielded 200,000 or more herring at a haul, and up to 10 or 15 years ago took probably 15,000 to 30,000 fish at a haul on an average. Only a few years back from 1,000 to 1,500 shad were frequently taken at one set of the seine. In 1913 the largest haul was 3,000 herring and 100 shad, while many times only 6 to 20 shad were taken.

In the same report, Smith attributed these declines he described for the Potomac River, halfway up from the Bay mouth, to a glut of fishing gear in the lower Bay:

Fish entering Chesapeake Bay have to run through such a maze of nets that the wonder is that any are able to reach their spawning grounds and deposit their eggs. The mouth of every important shad and herring stream in the Chesapeake Basin is literally clogged with nets that are set for the special purpose of intercepting every fish, whereas a proper regard for the future welfare of the fisheries and for the needs of the migrating schools would cause the nets to be set so as to insure the escape of a certain proportion of the spawning fish.

Smith wrote in 1917 that Virginia (which had first crack at the schools swimming in from the ocean) had deployed an astounding 16,793 gill nets and 2,012 pound nets in its waters in 1915. By comparison, Maryland (located farther up the Bay and thus dependent on fish not caught in Virginia's nets) set out 3,875 gill nets and 1,062 pound nets. The maze of fishing gear Smith mentioned in his 1914 report resulted from the understandable desire of seafood processors to be near shipping centers that could export processed fish to distant markets.[13] Thus, processors built their facilities near the Bay mouth and entrances to major rivers where incoming fish could be harvested before they dispersed upstream. Captured fish could then be processed promptly for shipment to market. As a result, since 1895 the greatest annual harvests came from the lower reaches of the Bay and its tributaries, starving upstream locations of fish.

Providing for free passage of migrating fish was one important way to maintain fisheries. Limiting harvests was another, and the one could supplement the other. Thus, Commissioner Smith wrote in 1914 that lowering the harvest by as little as 10% by limiting the number of nets might enable fish production to grow to sustain the fishery. But nothing happened until World War I intervened. Serendipitously, a decree by the War Department in 1916 required that navigational lanes among the maze of nets be kept open for military ship traffic in the lower Bay. Smith reported in 1918 that this requirement had had positive effects on the fishery, so Virginia's legislators passed a law in 1917 to allow more room in the Bay for fish to bypass nets.

Despite the various recommendations that dams be removed, pollution prevented, and fishing effort curtailed, harvests continued to decline as business kept on much as usual.[14] When John Snyder investigated opportunities for culturing fish on Maryland's Eastern Shore in 1917, he found that only about one-tenth as many shad were returning to local rivers compared with 15 or 20 years earlier. Whereas alewives had previously spawned in great numbers in the shallows of rivers, creeks, and rivulets, they were nearly extinct there in 1917. As for blueback herring that normally spawned in deeper waters than alewives, they were spawning in "very greatly diminished numbers" compared with years past. The baselines for harvests of shad and river herring in the Chesapeake Bay had shifted downward over the decades.

In an article in 2009, fisheries scientists Karin Limburg and John Waldman used historical data to demonstrate how the choice of a baseline informs restoration goals. In their figure 2 they showed that the baseline chosen by managers for contemporary shad restoration is the 1887 fishery (when the US Fish Commission began keeping records). Harvests then were about 14 to 15 million pounds of fish. However, earlier in 1832 an estimated 110 million pounds were captured. Today's goal is a fraction of that amount. There are undoubtedly good reasons for selecting a smaller target, but the point is that it is less than it could be.

This illustration of the relation of the baseline selected to restore shad to earlier baselines that might have been chosen raises the question of what data might provide a more realistic restoration goal. The goal of about 14 to 15 million pounds is proving difficult to reach. Does it make sense to aspire to attain the early harvests of about 110 million pounds? This is a difficult question to answer and will require information on how the ecology of the Bay and its tributaries has changed over time, how the present carrying capacity of the system compares with the past, and what human and financial resources are available to push restoration actions towards a higher baseline.[15]

Meanwhile, the depleted numbers of shad and river herring have resulted in Maryland and Virginia placing a moratorium on harvesting the fish in recent decades. Today, when the shadbush blooms in the spring, no seines, gill nets, or pound nets are deployed, and fishing floats on the Susquehanna are no more. Perhaps as dams are removed, pollution prevented, and spawning habitat restored, future shadbush blooms might coincide with the sight of bow waves generated by the upriver surge of renewed populations of shad and river herring swimming vigorously toward their spawning grounds.

The World's Greatest Oyster Fishery

An Expansion, Then a Crash

Oysters there be in whole bancks and bedds, and those of the best: I have seene some thirteen inches long. The savages use to boyle oysters and mussells togither, and with the broath they make a good spoone meat, thickned with the flower of their wheat; and yt is a great thrift and husbandry with them to hang the oysters upon strings (being shauld and dried) in the smoake, thereby to preserve them all the yeare.

Strachey (1610)

Here [in the Chesapeake Bay] are such plenty of oysters as they may load ships with them. At the mouth of Elizabeth River, when it is a low water, they appear in rocks [reefs] a foot above water.

Glover (1676)

The abundance of oysters is incredible. There are whole banks of them so that the ships must avoid them. A sloop, which was to land us at Kingscreek, struck an oyster bed, where we had to wait about two hours for the tide. They surpass those in England by far in size, indeed they are four times as large. I often cut them in two, before I could put them into my mouth.

Michel (1702)

MICROSCOPIC OYSTER LARVAE PRODUCE calcium carbonate shells that enclose their soft tissues while they develop in plankton for two or three weeks. When they are ready to leave the plankton, larvae search out hard surfaces, especially the shells of older oysters, and cement to the surfaces (figure 4.1), whereupon they are called spat. This action produces aggregated masses of oysters, called banks, beds, rocks, or reefs in the quotations above. Over time and without human harvesting, reefs build up oyster upon oyster, generation upon generation, live upon dead, providing solid

Figure 4.1. Oysters attached to a round boulder and to each other. The whiter surfaces are the interior of three dead oysters that have lost their upper shell. Brooks 1891.

structures over the bottom of the otherwise soft-bottomed estuary that organisms in addition to oysters can exploit as habitat (see figure 2.5). For this reason, scientists call oysters "ecosystem engineers."[1]

Throughout history, as demonstrated by prehistoric shell middens and by written accounts, oyster reefs have supported extensive fisheries worldwide. In the United States the nationwide demand for oysters in the 1800s was described by Dr. Thomas Nichols in 1864:

> The American oyster, from New York to New Orleans, is large, bland, sweet, luscious, capable of being fed and fattened, and cooked in many styles, and is eaten for breakfast, dinner, supper, and at all intermediate hours. Oysters are eaten raw, pure and simple, or with salt, pepper, oil, mustard, lemon-juice, or vinegar. At breakfast they are stewed, broiled, or fried. At dinner you have oyster-soup, oyster-sauce for the fish, fried oysters, scalloped oysters, oyster pies, and when the boiled turkey is cut into, it is found stuffed with oysters. Some of these oysters are so large that they require to be cut into three pieces before eating.

Nichols noted the abundance of "oyster cellars" (eateries) in New York and the animals' cheapness therein (all you could eat for sixpence, or an eighth of a Spanish dollar). Oyster suppers were also "as regular a thing in Cincinnati or St. Louis as in New York or Baltimore." Demand for cheap oysters continued, as revealed by George Makepeace Towle's description of their availability, the many ways they could be prepared, and their cost, in 1870:

The oyster may perhaps be called the national dish—it is at least the great dish of the Atlantic states. They are within the reach of every man, for they are cheap and plenteous . . . There is scarcely a square without several oyster-saloons; they are above ground and underground, in shanties and palaces. They are served in every imaginable style—escolloped, steamed, stewed, roasted "on the half shell," eaten raw with pepper and salt, devilled, baked in crumbs, cooked in pâtés, put in delicious sauces on fish and boiled mutton . . . The oyster is the sine qua non of all dinner parties and picnics, of all night revels and festive banquets. For tenpence you may have a large dish of them, done in any style you will, and as many as you can consume. The restaurants—ostentatious and humble—are in the season crowded with oyster lovers: ladies and gentlemen, workmen and seamstresses, resort to them in multitudes, and for a trifle may have a right royal feast.

Frank Leslie's *Popular Monthly* elaborated on the versatility of oysters as a food in 1886, printing "A Song of the Oyster," presumably to be sung heartily accompanied by strong drink:

Let us royster with the oyster—in the shorter days and moister,
That are brought by brown September, with its roguish final R;
For breakfast or for supper, on the under shell or upper,
Of dishes he's the daisy, and of shell-fish he's the star.
We try him as they fry him, and even as they pie him;
We are partial to him luscious on a roast;
We boil him and we broil him; we vinegar and oil him,
And oh! he is delicious stewed with toast.[2]

The demand for this inexpensive and adaptable food boosted oyster fisheries in the United States. New England dominated in the 1700s and early 1800s, but the Chesapeake Bay eventually became the leading fishery as New England overharvested its local stocks in the 1800s. Simple gear like tongs and dredges were used in the harvest (figure A.6). Here is an early description of tongs: "The inhabitants usually catch them on Saturday. It is not troublesome. A pair of wooden tongs is needed. Below they are wide, tipped with iron. At the time of the ebb they row to the beds and with the long tongs they reach down to the bottom. They pinch them together tightly and then pull or tear up that which has been seized" (Michel 1702).

Michel makes tonging sound easy, and it may have been so for fishers plucking oysters from shallow inshore reefs that extended high off the estuary bottom.[3] However, people who are not watermen have no idea how challenging it is to use tongs these days to haul oysters up from diminished reefs deep on the estuary floor (figure 4.2). Much of the oyster season (October to March in Maryland, April in Virginia) includes winter, so near-freezing or freezing temperatures are common. Watermen set out in the dark Monday through Friday when the weather allows (and it often does not). In the past, tongers sailed to a nearby oyster reef, but today they fire up the engine of

Figure 4.2. (*Top*) A tonger stands on the gunwale while tonging in deep water (note the long shafts on the tongs in the background). Another waterman culls the catch on board. (*Bottom*) Emptying the tongs on the culling board. Courtesy Skip Brown and Maryland Sea Grant College.

their small boat and motor out to begin tonging at first light. They drop heavy shaft tongs to the Bay bottom, scissoring the upper arms of the tongs back and forth to rake oyster reef material into the basketlike jaws of the tongs, which they then muscle to the surface and empty on a sorting board. There, live legal-sized (3 inches or more) oysters are culled, or extracted, from among the mud, shell, and smaller oysters that were collected by the tongs (figure 4.2). If the weather is rough or the gunwales covered in ice, tongers risk falling overboard, and most have not learned to swim. Even if they could, the cold water would rapidly bring on hypothermia. Here is a *New York Times* report from the James River in 1880 about the seamanship of tongers: "To see the oystermen balancing themselves in one of their canoes, and working with so much energy at the same time, was quite a novelty. Many of these canoes are so narrow that should a novice step into one it would most probably be overturned; yet the oystermen work in them all day long in smooth weather, and sometimes in pretty stormy weather, and apparently keep them properly balanced without any effort."[4]

Ernest Ingersoll wrote in 1887 that tonging involved great exposure and hard labor, with the injury to health from exposure being such that few tongers reached old age. Tongers using "patent tongs" were at less risk, not having to bend over or stand on the gunwales. However, they still had to keep an eye on the weather.

Dredgers were at somewhat less risk than tongers in that they did not stand on the gunwales. But dredges are heavy and difficult to handle, especially when full, and the booms of the sails can swing around and break a skull or knock an unwary dredger into the water. Any ice on a heaving deck in winter could lead to falls or even to a man overboard. In the 1800s, before winders (winches) became gasoline powered, one to four men cranked the winders to raise a loaded dredge from the estuary floor (figure 4.3). On occasion the dredge might hang up on an obstruction on the bottom, causing the winder handles to suddenly reverse and shatter a dredger's bones. One particularly terrible accident happened to a 45-year-old man in 1882 when a dredge became hung up and the crank handle suddenly hit him on the abdomen, hurling him against a hatch and fracturing and dislocating his spine.[5]

Charles Stevenson summarized 172 surgical cases involving accidents on dredge boats that were treated by Surgeon-General William Wyman at the US Marine Hospital Service in Baltimore during the winters of 1882–1883 and 1883–1884.[6] Fractures (and the spinal dislocation mentioned above) were caused by 21 malfunctions of crank handles of winders, 14 falls on slippery decks, and contact with 7 "foreign bodies," resulting in 72 broken bones. Contusions and lacerations were caused by 19 foreign bodies, 16 falls, and 15 errant crank handles. There were 50 frozen extremities. Thirty men suffered from "oyster-shell hand," a severe bacteria-induced inflammation and swelling of the hand as a result of cuts from handling oyster shells; bones and tendons might become exposed, and fingers might be lost. Wyman's numbers did not include similar surgical cases in the other Maryland ports of Crisfield, Cambridge, Oxford, and Annapolis or in Virginia ports, nor did they include incidents of

Figure 4.3. Men operating two-handed winders to set and retrieve oyster dredges at the stern of a dredge boat in the late 1800s. The empty dredge on the left is going overboard from the winder obscured by the man whose back is turned; the upper frame of the dredge on the right can just be seen. Courtesy Calvert Marine Museum, Solomons, Maryland.

"pneumonia, pleurisy, and rheumatism" brought on by the inadequate clothing of the men and unhygienic conditions in the cramped living quarters on board dredge boats.

But oysters were bounteous and in high demand, thus encouraging oystering in spite of the dangers involved. The industry grew rapidly to exploit the abundant bivalve, whose numbers when Europeans arrived amazed observers (see the quotes at the head of this chapter). The environmental reasons thought to support this bonanza were explained by an anonymous observer in 1869:

> The Chesapeake Bay and its tributaries afford the most favorable conditions for the natural growth of the oyster, as well as all needful facilities for its artificial propagation and culture. Located in the proper temperature, its bottoms of sand and rock, its abundant produce of sea-moss as a home and breeding place, its waters tempered in degrees of saltness to suit all varieties, and its numerous fresh-water streams, bringing down in their floods a continuous supply of food and other requirements, render the bay superior, in its oyster grounds, to any body of water on this continent or perhaps in the world.

The extensive shoals flanking the Bay's channel provided space for oyster reefs to thrive (figure 4.4) as the individual oysters fed on the phytoplankton that grew vigorously in the sunlit waters.[7] Scattered shallow-water reefs suitable for tonging extended up to the Bush River in the upper Bay, which is no longer true today, as lower salinity and silt have made that environment unsuitable. Oysters also live in deeper channels (figure 4.4) where dredges are needed to harvest them. Because pre-Colonial oysters had been left undisturbed to grow (except for the presumably modest amounts harvested from shallows by Indians), they reached larger sizes than they do at present; shells 14 inches long have been found in middens.[8]

With minimal harvesting occurring after Europeans arrived, oyster abundances continued to impress observers as late as two centuries after the accounts at the head of this chapter: "From Kent Island, within twenty-five miles of Baltimore, to Cape Henry, a distance of one hundred and forty miles, the bottom of the bay is, with slight exceptions, a continuous oyster bed [see figure 4.4]. All the freshwater streams that empty into the bay within the above-named limits are stocked, either naturally or artificially, with oysters, as far up toward their sources as the influence of the salt water extends" (Anonymous 1869). That abundance and geographical extent of oysters in the Bay (the baselines) were about to change as demand for oysters accelerated and as national and international transport systems improved.[9]

Environmental Effects of Tonging and Dredging

Tonging by the thousands of watermen working in the 1800s removed oysters from oyster reefs and redistributed shell material in their vicinity, gradually lowering their profile. However, tonging is so labor-intensive that it probably took decades for profiles to be lowered greatly. In 1910, scientist Frank Moore measured the amount of bottom that could be grabbed by tongs of different lengths in waters of different depths in the James River. He showed (unsurprisingly) that in deeper water more work is involved and fewer grabs can be made in a given time period. An oysterman using 14-foot tongs could cover twice as much bottom per grab in 4 feet of water as he could in 8 feet of water and about two and a half times as much bottom compared with using 20-foot tongs in 16 feet of water. Thus, a tonger could take about 2.7 grabs per minute in 4 feet of water, 2.6 in 8 feet, and 1.6 in 16 feet. Of course, when oysters were plentiful, each grab would bring many oysters to the surface. As oysters became depleted, grabs brought up fewer oysters and much more shell, increasing the labor needed to cull enough legal-sized oysters from the shell to fill a bushel (in Maryland this was 329 oysters on average in the late 1800s).[10]

The more efficient mechanical dredging from sailboats undoubtedly took less time to have a major effect on the shape and extent of deeper oyster reefs than did tonging. A dredge boat capable of holding a thousand bushels usually deployed two dredges measuring 5 feet wide across the mouth, one dredge over each side. On a daily basis a dredge boat might travel at least 30 miles while sailing back and forth over an oyster

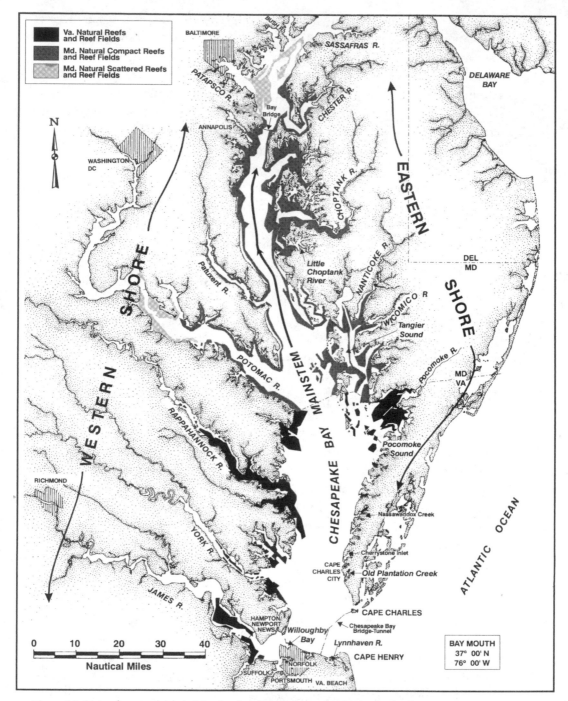

Figure 4.4. Natural oyster beds in Maryland and Virginia as of 1894. Luckenbach, M., R. Mann, and J.A. Wesson. (eds.), 1999. Oyster reef habitat restoration. A synopsis and synthesis of approaches. Virginia Institute of Marine Science Press, Gloucester Point, VA.

reef, thus mining about eight acres of bottom.[11] In 1869, an anonymous author wrote that there were 1,700 dredge boats licensed to fish in Maryland, which gives an idea of their potential cumulative effect on oyster bottoms.

Although dredging, like tonging, ultimately removed huge quantities of oysters and shell, greatly lowering the vertical profile of deeper Bay reefs, when it first began it had the beneficial effect of ripping apart oyster clumps (figure 4.1) and scattering the smaller clumps and individuals over a wider area of bottom, so that: "A half century since [ago], the bottom of the Chesapeake was interspersed with numerous isolated beds of small extent and great thickness, but dredging has so scattered them that they now form almost a continuous bed covering the whole bottom" (Anonymous 1869).

In 1881, Ernest Ingersoll described how dredging changed oyster reef distributions in the early years of exploitation:

> In a report upon the 'oyster-beds of the Chesapeake bay', made in 1872, by Mr. O. A. Brown, to the auditor of public accounts of Virginia, it is said that "The dredging of oysters is as necessary to their development and propagation, as plowing is to the growth of corn; the teeth of the dredge take hold of the rank growth of the oyster-beds, and, by being dragged through them, loosen them . . . and gives them room to grow and mature properly; moreover, beds are continually increased in size, for when the vessel runs off the rock [oyster bed] with the chain-bags filled with oysters, the oysters are dragged off on ground where no oysters existed, and thus the beds are extended; and when the vessel is wearing or tacking to get back on the oyster-beds, the catch just taken up is being culled off, the cullings thrown overboard to form new cultch for drifting [larvae] to adhere to. Reliable oystermen tell me, that since dredging has been carried on in Tangier and Pocomoke [Sounds] the beds have more than doubled in size, and, with the moderate force that worked upon them prior to the [Civil] war, were continually improving.

However, Ingersoll also warned about the negative effects of unrestricted dredging:

> As the best-stocked and most productive beds of Europe were quickly destroyed by unrestricted dredging, so may the hitherto seemingly exhaustless beds of the Chesapeake Bay be depleted, if the present rate of dredging is continued. An illustration of this may be seen in the almost total exhaustion of the once famous beds of Tangier and Pocomoke Sounds. Year after year these beds were dredged by hundreds of vessels, and even the summer-months afforded them but little rest. The result of this has been plainly seen during the past few years, and more especially during the season of 1879–80, in the great scarcity of oysters in these sounds.

Some years before Ingersoll's warning, some (but not all)[12] observers were concerned:

> The oyster commissioners of the Chesapeake report a gradual diminution in the oyster crop in the past ten years, and estimate, by the same rates of decrease, that

the whole stock will be exhausted in a half century . . . Until recently the supply by natural increase was considered inexhaustible, and no aids, either legislative or otherwise, were deemed needful or advisable. But now, when an interest of so much importance to the States most directly concerned, and to the whole country, is threatened with extinction, the means for its preservation become a necessity . . . by proper management, the oyster grounds of the Chesapeake can be made to supply a demand equal to that of our whole country at the present time. (Anonymous 1869)

Rise and Fall of the Oyster Fishery

Observers were right to be concerned about possible overfishing of oysters in the Bay. They knew that overfishing in New England drove harvest fleets south along the East Coast, as will be described below. Nevertheless, for a long period the mining of Maryland's oyster reefs supported the livelihoods of thousands of individuals.[13] "In the city of Baltimore seventy houses are engaged in the oyster business, mostly in canning for exportation, while at various points in the bay are establishments that employ from fifty to four hundred hands each, during the season, in opening and canning. By the official reports there are fifteen thousand persons engaged in the business of oyster fishing and a fleet of one thousand seven hundred vessels of fifty tons burden, and over three thousand smaller crafts, are duly licensed for the trade" (Anonymous 1869).

The 1,700 larger vessels mentioned would be dredgers, and the 3,000 smaller crafts would be predominantly tongers, with some being scrapers using a smaller dredge. A photo (figure 4.5) taken in the late 1930s when oyster stocks were much depleted compared to the nineteenth century shows about 60 dredgers sailing over an oyster reef, giving some idea of the fishing intensity of 1,700 dredgers. Only about 30 or 40 dredge boats are still afloat today, with few of them operating.[14] In recent years, about 1,000 people have oyster licenses in Maryland, a decline from the numbers mentioned above. Further, not all license holders are actively oystering; participation by some depends on how productive the oyster season seems to be.

The numbers of persons in the oyster industry continued to rise after 1869. In 1875, Edward King wrote: "The oyster trade of Baltimore is stupendous. Whole streets are devoted to the packing of oysters; and twenty thousand men, women, and children, are employed either in fishing them up, or packing them down. From the vast waters of the Chesapeake many persons have already wrested handsome fortunes . . . In one single establishment in Baltimore, fifty thousand cans of raw oysters are packed each day. The manufacture of tin cans is itself a giant business; and several large printing houses are constantly occupied in printing labels."

In 1891, 32,104 Maryland residents and 18,593 Virginia residents were employed in the oyster industry, for a total of 50,697 for the Bay as a whole (table 4.1; Smith 1895). By comparison, in 2014, 22,950 people were employed in the combined sea-

Figure 4.5. Aerial view of skipjacks and bugeyes dredging south of Black Walnut Point, Tilghman Island, Maryland, in the late 1930s. Photograph from the H. Robins Hollyday Collection at the Talbot Historical Society, Easton, Maryland.

Table 4.1 Number of persons working in the 1891 Chesapeake Bay oyster fishery

State	Dredging	Tonging	Transporting	Processing, marketing	Total
Maryland	6,862	12,505	1,444	11,293	32,104
Virginia	3,221	12,421	701	2,250	18,593
Total	10,083	24,926	2,145	13,543	50,697

Source: Smith 1895b.

food industries (not just oysters) of Virginia and Maryland (see table 1.1), with 7,815 being seafood harvesters.[15]

The early history of Maryland's fishery from 1839 until 1911 was displayed in an annotated graph prepared in 1912 by Dr. Caswell Grave of Johns Hopkins University (figure 4.6).[16] His data points were estimates based on records from oyster packers and information on the number of oyster licenses issued, etc., so their accuracy might be questioned. However, in 1937, A. J. Nichol carefully compared processing statistics assembled by the Baltimore Board of Trade with the 1865 records of Baltimore canner Caleb Maltby and with the 1868 records compiled by the US Fish Commission; he found the records to be similar. In 1964, Dr. Chris Christy discussed the

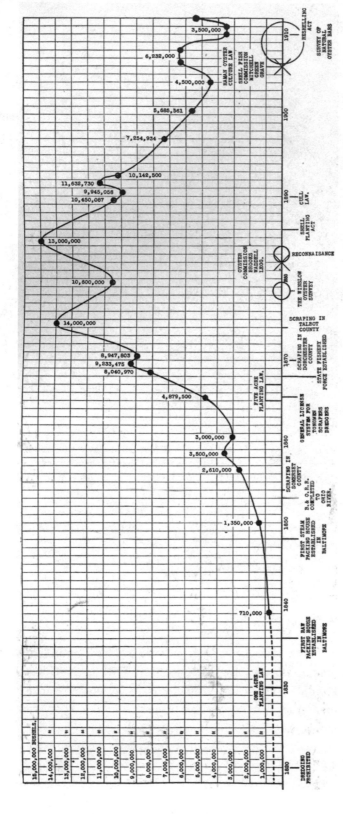

Figure 4.6. Landings of oysters in Maryland from 1839 to 1912 in millions of bushels, with key events in the history of management noted. Grave 1912. Courtesy Calvert Marine Museum, Solomons, Maryland.

dependability of these early harvest estimates. He concluded that early harvests were probably not greatly overestimated and that the decline was probably due to over-harvesting and possible increased siltation of oyster reefs caused by land clearing (excessive silt hinders settlement by larval oysters, which require a clean surface for successful settlement).

While the data on harvests in figure 4.6 are probably reasonably dependable, note also that an anonymous statement in the 1869 report of the US Commissioner of Agriculture indicated that the estimated millions of bushels taken from the Bay in his day did not include "oysters taken from private beds or plantations, owned by the residents on the islands and the shores of the bays and rivers, who do not regularly engage in the trade, but cultivate them for their own uses; nor the numbers taken by the half-piratical 'pongys,' canoes, and other small craft that continually depredate upon the beds without the required license."[17]

Subsequent estimates of harvests were perhaps more quantitative than those of Grave and now include those of Virginia (figure 4.7). The decline over time is clear.

In the late 1800s, the Chesapeake Bay had become the greatest oyster-producing region of the world (table 4.2). In 1891–1892 the combined harvests of Maryland and Virginia made up 59% of the total volume of US oyster harvests and 50% of the value of oysters nationally. And 20% of Americans in the US fishing industry were involved in the Bay industry. Virginia's harvest was about equal to that of the combined harvests of all foreign oyster-producing nations, and Maryland's harvest was twice that amount.[18]

The factors that led to this dominance by Maryland and Virginia should have provided an object lesson about the effects of overfishing; sadly, the two states ignored the lesson for some time. Oystering in the Bay increased because New England's oyster reefs had become badly overfished throughout the early nineteenth century.[19] A major player in the early US oyster industry had been Connecticut, especially Fair Haven, the nation's first oyster-packing center.[20] Beginning in about 1808, as their local reefs became less productive, oystering firms in New England sent dredge schooners to New Jersey and Virginia to harvest oysters for planting on New England's depleted reefs.[21] Concerned about this incursion, Virginia legislators in 1811 banned dredging in Virginia waters. This forced the northern fleet up the Bay into Maryland waters, whereupon Maryland legislators passed a law in 1820 banning dredging as well as the transport of local oysters in ships not locally owned for the previous year by Maryland residents.[22]

These laws led to Bay residents becoming more involved in oystering, both to provide oysters to New England and for local use. As the Bay fishery grew, legislation was passed to govern harvest practices.[23] The rich grounds in Tangier Sound had not yet been discovered, so there was concern that the shallow-water oyster reefs could be overfished. Modest protective laws followed, such as ones banning burning oysters for use as agricultural lime in Dorchester and St. Mary's Counties (1836) and Somerset County (1840).

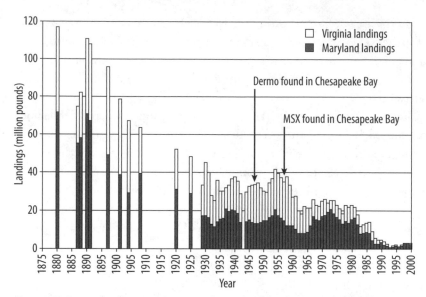

Figure 4.7. Oyster landings in Virginia and Maryland, beginning in 1880 and declining thereafter. Reprinted from *Nonnative Oysters in the Chesapeake Bay* (2004) with permission from the National Academies Press, Washington, DC.

Table 4.2 Harvest quantities and values for the top five oyster-producing US states and for the top three oyster-producing foreign countries in 1891 or 1892

State or country	Number of bushels harvested	Value ($)	Year
Maryland	11,633,000 (39%)	5,866,000 (35%)	1892
Virginia	5,985,000 (20%)	2,488,000 (15%)	1891
New Jersey	2,632,000 (9%)	1,747,000 (10%)	1892
New York	2,611,000 (9%)	2,749,000 (17%)	1891
Connecticut	1,940,000 (7%)	1,426,000 (9%)	1892
Chesapeake Bay (VA + MD)	17,618,000	8,354,0000	
All United States total	29,796,000	16,639,000	
British Isles	2,760,000	6,200,000	1891
France	2,000,000	5,000,000	1891
Canada	153,000	184,000	1891
All foreign countries total	5,461,000	12,699,000	

Source: Modified from Stevenson 1894.

Note: Values within parentheses are the percentage of the total US harvest and landed value for that time period.

In 1865, Maryland passed a General License Law that abolished earlier oystering laws and enacted a new code that established a statewide license system governing tongers, scrapers, and dredgers and that allowed large dredges to be used for the first time since 1820. This opened waters deeper than 23 feet (generally out of the reach of tongers) to dredging to help meet the rising demand for oysters. Steamboats or machinery for harvesting were banned as too efficient, so dredge boats used sail for propulsion and human-operated winders to deploy and retrieve dredges. Dredgers faced a closed season from June 1 to September 1 (the oyster's reproductive season); tongers had no closed season. Many watermen ignored the law, so in 1868 Maryland established a State Fishery Force ("oyster navy" or "oyster police") to patrol Maryland's portion of the Bay and its tributaries.[24] Virginia established a similar police force in 1875.[25]

With dredgers exploiting deeper waters and tongers working the shallows, harvests expanded greatly (figure 4.6): in 1875 about 14 million bushels of oysters were landed in Maryland.[26] A subsequent decline led a concerned Maryland General Assembly to commission Lieutenant Francis Winslow of the US Coast and Geodetic Survey to examine the extensive oyster grounds in Tangier and Pocomoke Sounds in 1878 and 1879.[27] The Pocomoke Sound survey found differences in the structure and biology of grounds never before fished compared with older heavily fished grounds (table 4.3). Dredging on oyster reefs broke up clumps of oysters, dragging smaller clumps and individual oysters and shell onto new bottom and thus expanding the area of the reefs. These oysters were now less crowded and presumably had easier access to their planktonic food, not having to compete so much with attached neighbors. Instead of being long and thin like most oysters living in crowded clumps on unfished beds, oysters on fished reefs were more rounded and could grow to a larger size and be in better condition as measured by their plumpness. Also, the unfished reefs contained a smaller proportion of shell debris (about 30%) than did the fished reefs (shell proportions measured up to 97% in some heavily harvested parts of Pocomoke Sound).

Attempts to quantify the effects of dredging led to the following data reported by Winslow in 1884: in 1876, Chesapeake Bay oyster reefs yielded an average ratio of 3.7 bushels of oysters to 1 bushel of shells, compared with diminished ratios of 1.9 to 1 in 1879 and 1.3 to 1 in 1882. In Tangier Sound there was an average of 1 oyster for every 2.3 square yards in 1878–1879 and 1 for every 4.2 square yards in 1883. Oyster numbers in these regions were clearly declining rapidly.

Based on his surveys, Winslow proposed in 1881 that a commission be established, free of political interference, to make management recommendations. His own enlightened recommendations included controlling destructive over-dredging, preventing harvest of young oysters by setting aside reserves with abundant young, adding cultch (oyster shell, broken pottery, etc.) to depleted reefs as material for oyster larvae to attach to, and controlling predators and pests. In his 1884 report he urged the legislature to allow private oyster culture on Bay bottom leased from the

Table 4.3 Characteristics of unfished and fished oyster beds in Pocomoke Sound

Unfished	Fished
Oysters in clumps of 3 to 15	Single oysters, or in clumps of 2 or 3
Bed with minimal amounts of sand or mud	Beds with much sand or mud among shells
Clean shells free of worms	Shells infested with worms or bore holes and much broken
Oyster shells long and narrow with thin, sharp bills	Oyster shells large and broad with thick bills
Soft tissue long and thin	Soft tissue plump
Broken shell and debris made up ~30% of material dredged from the beds	Broken shell and debris made up to 97% of material dredged from the beds

Source: Winslow 1881.

state. His studies had shown that the oyster yield in Maryland's public fishery was 40 bushels per acre compared with triple this yield in New England states where private culture was performed on less acreage.

In 1882 the state legislature followed Winslow's advice and established an Oyster Commission headed by Johns Hopkins University Professor William Brooks. The commission's task was to examine the oyster reefs and to advise as to their protection and improvement. Although the legislature provided no financial support for the work of the commission, the governor provided some money from his emergency fund, and the university granted Brooks two years of paid leave to perform the survey. The commission provided a report based on sampling 59 oyster reefs, making 326 dredge hauls covering 121,000 square yards of oyster bottom, and capturing and examining more than 30,000 individual oysters.[28] As had Winslow, the commission noted a decline in numbers of young and adult oysters and an increase in the proportion of shell to live oysters. It again encouraged leasing of oyster bottom to allow private oyster cultivation and scientific surveys to determine the extent of oyster reefs.

Professor Brooks turned the Oyster Commission's report into a book for popular consumption that is still available and enlightening.[29] In it, he proposed reasons for the decline in both larval settlement and subsequent numbers of spat and older oysters—that numbers of mature oysters available to furnish spawn had declined markedly; that spat attached to shell were killed by watermen and processors who ignored culling laws designed to have spat thrown back on their natal oyster bed; and that there was a lack of clean shell for cultch because of its diversion to the road building, agricultural lime production, and chicken grit industries. He described the success of private oyster culture in Europe and New England, urging that Maryland support similar culture activities on leased Bay bottom to expand the Maryland fishery. Leasing would allow for placing adult oysters on barren bottom where they would spawn. Lease holders could also spread cultch that larvae could settle on and grow, thereby forming new oyster reefs.

In the year after the Oyster Commission's report appeared, the 1884–1885 oyster season saw over 15 million bushels of oysters harvested in Maryland, apparently due to an excellent oyster set in 1883.[30] This bonanza undercut the commission's recommendations, so they were not implemented.[31]

Most Maryland oystermen in the late 1800s were strongly against private culture actions that involved leasing Bay bottom (Virginia oystermen were more supportive).[32] In their 1916 report, Shell Fish Commissioners Green, Revell, and Maltbie attributed the Maryland oystermen's negative attitudes about leasing to a number of concerns. One was that the powerful canning industry could monopolize private culture, stockpiling oysters purchased cheaply in spring and then employing house employees to harvest them, thus converting oystermen from freelance, self-sufficient, and independent workers to wage earners working for the industry. A second concern was based on a widespread belief that cultivating oysters on barren bottom was impossible. By this reasoning, if an area were to yield oysters as a result of culture practices, it must have been an unrecognized natural oyster bed and should not have been privately leased (and thus made inaccessible to oystermen) in the first place.

In spite of this resistance to the commission's recommendation on leasing, some useful laws did get passed. In 1890, a cull law required that shells with attached spat as well as oysters smaller than 3 inches be thrown back onto the bed from which they were harvested so that they could continue to grow. Maryland was the first state to establish such a law.[33] It was unpopular with oystermen, who had been selling small oysters to steam canners or to private oyster growers from out of state for use as seed oysters, and many ignored it.[34]

Despite cull laws and other restrictions, the commercial harvest continued to decline (figures 4.6 and 4.7) as oyster reefs were mined in response to increased demand from the rest of the nation and thanks to the expansion of roads, canals, steamship lines, and railroads that allowed for rapid transport of whole and canned oysters. Maryland's oyster canning industry peaked in about 1885, declining thereafter as harvests fell and as the demand for fresh, not canned, oysters grew.[35]

Concerned about the continuing decline in Maryland, Professor Brooks wrote in the second edition of his seminal book in 1905: "the oyster grounds of Virginia and North Carolina, and those of Georgia and Louisiana, are increasing in value, and many of our packing houses are being moved to the south, but there is no oyster farming in Maryland, and our oyster beds are still in a state of nature, affording a scanty and precarious livelihood to those who depend on them."

The arguments over leasing in Maryland in subsequent decades pitted agricultural and manufacturing interests against oystermen because increased tax revenues that were expected to be derived from revitalized oyster reefs on leased bottom were to be used to improve public roads and bridges.[36] Newspapers wrote about the matter.[37] Between May 12, 1905, and April 11, 1906, the *Baltimore Sun* published at least 38 editorials about leasing and oyster culture and criticized the General Assembly's handling of the matter. An editorial on June 17, 1905, entitled "The Oyster and the

Politician," noted the public support for leasing by most merchants and farmers and the increasing sentiment among some (not many) oystermen for private culture.

In 1906 an oyster bill did pass (the "Haman Bill") that allowed individuals to lease up to 30 acres of barren bottom in county waters and up to 100 acres in the Bay beyond county limits.[38] A Shell Fish Commission was formed to perform a six-year survey of Maryland's oyster bottom (the "Yates Survey" led by Charles Yates) in order to separate working reefs from barren bottom (Virginia had commissioned a similar extensive survey led by James Baylor a decade earlier).[39] The survey occupied 11,006 oyster investigation stations, made 159,530 soundings of the bottom, and plotted 8,600 hydrographic positions, producing 17 official documents and 43 large-scale charts for a total of 2,400 printed pages and 400 square feet of charts.

Yates's survey data supplemented the earlier survey of the 1882–1884 Oyster Commission and the Winslow survey of 1878–1879. Further, the economic, historical, and social aspects of the fishery had been extensively described by Ingersoll and Stevenson.[40] This enormous accumulation of information, although incomplete in some aspects of oyster biology, should have been sufficient to help managers arrest the decline in production and to restore some of the former economic strength of the industry. At the end of his survey, Yates felt optimistic:

> It now seems not only reasonable but probable that within the next generation the citizens of Maryland will be leasing and cultivating a probable 100,000 and a possible 300,000 acres of so-called "barren bottoms" where oysters do not grow in commercial quantities: that more than 200,000 acres of natural oyster bars now reserved for the use of the oysterman as a result of the Maryland Oyster Survey will be so conserved and developed that they will produce, as they have done before, twice the amount they now yield; and that the oyster industry will then be based on annual production of 20,000,000 bushels of oysters where now it is barely 5,000,000.

However, it was not to be, as shown in figure 4.7 and detailed by Kennedy and Breisch and by Keiner. Resistance by oystermen and state legislators proved to be too strong until the early twenty-first century, and the harvest continued to dwindle to its present level of a few hundred thousand bushels.[41] This decline affected the employment of people beyond the harvesters. Thus Baltimore's canning industry alone employed 6,627 workers in 1880 and 8,687 in 1890 when harvests were highest, but only 900 in 1936–1937 when harvests had declined.[42]

Charles Stevenson predicted in 1894 that Maryland would be the last state to abandon the common fishery on public reefs, and he was right. The 1906 legislative session that had passed Haman's leasing bill and established the regulatory Shell Fish Commission had been argumentative, and the 1914 session was equally contentious. Oystermen brought pressure on their representatives to repeal the Haman Bill.[43] Controversy ensued. The *Baltimore Sun* printed at least 33 editorials between January 24 and April 11, 1914, defending private oyster culture. In March 1914, the newspaper printed two pages of comments by 99 Maryland citizens protesting the repeal legislation (the "Shepherd Bill"). Nevertheless, the bill passed. Supporters stated that it was intended to supplement the Haman Bill, but it actually restricted oyster culture. It provided that any Bay bottom that had been fished by any oysterman once in five years, even if only for a day, was natural, not barren, bottom and could not be leased. Three or more Maryland residents could dispute a barren bottom designation in circuit court, bypassing the Shell Fish Commissioners responsible for classifying bottom as barren. Almost immediately the courts removed 54,000 acres from the barren bottom category as a result of legal challenges, and between 1915 and 1963 an additional 15,000 acres were reclassified as natural bottom.[44]

Although the amount of remaining leased bottom was minimal, it yielded a relatively high harvest of oysters compared to the yield on public grounds. Thus, in 1948, 12,000 acres of bottom being used by state or private interests produced over 700,000 bushels of oysters, or about 60 bushels per acre; in that same year the 260,000 acres of public grounds yielded 1.4 million bushels for an average yield of about 6 bushels an acre. The decline on the public reefs was further illustrated by the fact that, from 1870 to 1890, over 1,000 dredge boats landed an average of 50 bushels of oysters per acre of dredging ground; the 1947 harvest by the 48 dredge boats still fishing averaged 1 bushel per acre.[45]

The declining fishery in Maryland continued to be examined by interested bodies seeking solutions to the downturn. In the 1900s, at least five organizations or legislative committees made useful recommendations, but these continued to be ignored or weakly implemented.[46] Unfortunately, political considerations rather than limited biological knowledge have frequently contributed to declines in fisheries in North America and elsewhere, and Maryland proved no exception.[47] Thus, Maryland's oyster harvest is a fraction of what it probably could be (it amounted to about 225,000 bushels in 2016–2017). The harvest decline was accelerated by the onset of two diseases—MSX and Dermo—that killed many oysters before they reached harvest size (figure 4.7).[48]

It is only in the past decade that aquaculture has been given a boost by Maryland's legislature, with encouraging results.[49] With strong actions underway to rehabilitate and restore oyster reefs in Virginia and Maryland, the wild fishery and aquaculture may eventually yield millions of bushels of oysters rather than the few hundred thousand harvested now. As the baseline moves back upward, perhaps in time folks will once again "royster with the oyster" as enthusiasts did 130 years ago.

Diamond-backed Terrapins

From Pig Food to Gourmet Delight to Protected Species

~~~~~~~~~~~~~~~~~~~~~

The waters [of the Chesapeake Bay] and especially the tributaries are
filled with turtles. They show themselves in large numbers when it is
warm. Then they come to the land or climb up on pieces of wood or trees
lying in the water. When one travels in a ship, their heads can be seen
everywhere coming out of the water.

Michel (1702)

BRIEF REFERENCES TO TURTLES by European explorers did not distinguish
among species, but Michel's report (above) of the Bay and its tributaries being filled
with turtles undoubtedly involves the estuary-dwelling diamond-backed terrapin
(figure 5.1). Turtles have been eaten by humans over the centuries around the world,
so it is not surprising that Indian kitchen middens excavated on the US Atlantic coast
have contained turtle remains, including those of terrapins.[1]

Although European colonists probably also used terrapins for food, historical accounts up to the nineteenth century reveal that the animal was very abundant but
apparently with no market value. For example, elderly residents of Maryland's Eastern Shore could remember when terrapins were fed to pigs.[2] Thus, terrapins were
probably just a cheap, local food for people living along estuarine shores. Anecdotes
described how former US Secretary of State John M. Clayton (born 1796) would oc-
casionally spend one or two dollars for an oxcart load of terrapins that were shov-
eled alive into his cellar in Delaware to be eaten over time.[3]

In the later nineteenth century, terrapins gradually became a commercial food and
an important item on the menus of hotels and gourmands. The New York Public Library collection of menus has "Maryland Terrapin" or "Baltimore Terrapin" appearing on over 160 hotel menus printed between 1887 to 1917, with hundreds of

Fig. 5.1. Diamond-backed terrapin.
Courtesy Willem Roosenburg.

additional terrapin-based dishes available from 1866 through 1958. And it was not just hotels. In the 1890s, the Old Bay Line of steamboats sailing up and down the Chesapeake Bay had an à la carte dollar menu offering terrapins (and canvasback ducks as well as Mobjack Bay oysters).[4] Not to be outdone, in 1900 the Baltimore and Ohio Railroad included terrapins (along with Lynnhaven oysters and canvasback duck) on its dining car dollar menu (figure 5.2).

Commercial fishing of terrapins initially centered in the Chesapeake Bay and North Carolina. Chesapeake animals grew larger than those in North Carolina and had a better reputation for taste. Potomac River animals were said to be the finest available, a point disputed in print by an admirer of Chester River animals.[5] The fishery showed the usual boom and bust pattern of other Bay fisheries (figure 5.3). As with other aquatic foods, some terrapins were probably always being captured for personal use and so were not included in commercial harvests.

Harvest data for all US states gathered by the federal government show a further decline from 1950 to the twenty-first century. However, beginning in 2002, US exports of terrapins began to increase, correlated with an increasing demand for turtle flesh from Asian markets, stimulating Maryland's harvest ban on terrapins in 2007.[6]

Terrapins have been harvested by a variety of simple techniques and devices. As autumn temperatures drop, terrapins bury into marsh sediments. Experienced hunters ("proggers" in the late 1800s) would wade the marsh looking for telltale mounds before thrusting a pole into the center of the mound and prying out whatever hard object was encountered, storing captured terrapins in a sack.[7] Terrapins buried under water too deep to wade through could be captured in modified oyster dredges or "drags," with terrapins captured by dredging in Somerset County, Maryland, as early as 1837.[8] Most drags had a coarse, large-meshed bag attached to a rectangular iron frame about 3 feet wide and with 3-inch teeth on the bottom; this was used to dig into the mud and root out the somnolent reptile. A dredge was more effective in colder

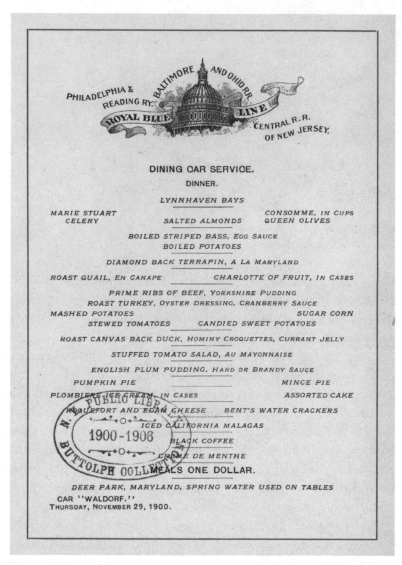

Figure 5.2. The dollar menu from the Baltimore and Ohio Railroad in 1900, offering oysters, terrapin, and canvasback ducks. New York Public Library menu collection.

weather when the terrapins were less active and unable to swim away quickly from the oncoming device. In warmer weather, a hungry animal might be captured by a simple baited hook (figure 5.4, top) and dip nets captured inquisitive individuals that popped their heads out of the water when the side of a boat was rapped.[9] A coarse-meshed seine (figure 5.4, bottom) was effective in colder weather in deep-water holes where terrapins had gathered.[10] One end of a seine, some measuring 400 or 500 feet long and 18 to 20 feet wide, was tied to a stake, with the free end run off a boat rowed in a circle back to the stake. The two ends were then closed on each other and the

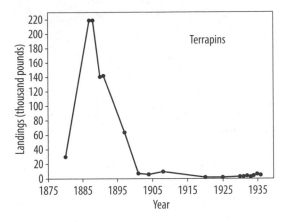

Figure 5.3. Diamond-backed terrapin landings in Maryland from 1880 to 1936. Data from McCauley 1945.

Figure 5.4. Catching terrapins with a baited line *(top)* and a net *(bottom)* near Annapolis, Maryland. *Frank Leslie's Illustrated Newspaper,* September 13, 1879.

seine quickly hauled into the boat, with the fishers rapping the boat bottom or sides with an oar to stimulate the inquisitive terrapins to swim off the bottom and become trapped. Finally, traps called fyke nets (figure A.7) came into use.

Hunters preferred capturing females because they grew larger than males and often contained eggs, considered a delicacy. Because females crawled onto beaches to lay their eggs, they could be captured by arranging boards on edge along the tide line to steer them toward shallow boxes sunk at intervals in the sand, in which they became trapped.[11]

Frederick True wrote in 1887 that that the first commercial market for terrapins apparently occurred in 1849 when Captain John Etheridge dredged 2,150 terrapins from Roanoke Sound in North Carolina and sold them in Norfolk for $400 (about $11,700 in 2016 dollars).[12] Inspired by his success, Etheridge dredged 1,900 more animals and sold them in Baltimore for $350 (about $10,250 in 2016). Other North Carolina residents copied him, with some shrewd individuals holding terrapins in enclosed semiaquatic pounds in the region for future sales locally and to Baltimore, Philadelphia, and New York. The market developed slowly until after the Civil War, when by 1877 live terrapins were shipped from Baltimore to Washington, Philadelphia, or New York from September to March.[13] By 1886 Baltimore was the greatest terrapin market in the world, with over 500 harvesters in the Bay collecting an estimated 600,000 terrapins said to be worth $1,500,000 (about $41 million in 2016).[14]

Although terrapin meat could be canned (figure 5.5), live animals were preferred. Since the greatest demand was in January and February, animals captured earlier were often kept in pens in cool, dark places like cellars. Terrapins stop eating in cold conditions, so their care and maintenance was minimal, although sometimes they were fed oysters or crabs in autumn months (while expensive today, oysters and crabs—especially peelers that died while being held in floats to shed their carapace as discussed in chapter 8—were very abundant and inexpensive in the nineteenth century).[15] However, in the early years of the market, prices were lower for terrapins that had been held in cellars or pens. Penned animals held on hard floors tried to dig through the substrate, wearing away their toes or developing corns on their feet and scratches on their under-shell.[16] Such evidence of captivity lowered their price.

Processors shipping live animals to domestic markets would pack terrapins surrounded by "seaweed" (or sea grass) in a barrel (figure 5.6) or wire basket before shipping them without food or ice; most animals arrived unscathed. Overseas shipments in winter went almost exclusively to England and France, with terrapins packed in hay or straw in barrels; nearly all would survive.

The price for terrapins rose as demand grew and as abundances fell, with the largest animals being the most expensive.[17] In 1877, animals measuring 7 inches or longer on their under-shell sold for $25 to $36 a dozen (about $572 to $824 in 2016 dollars) in Baltimore. In 1884, a Washington caterer noted that he was paying from $40 to $50 for a dozen large animals whereas in the 1860s he had paid $6 for a dozen.

Figure 5.5. Canned terrapin meat from Holland Terrapin Farms, Crisfield, Maryland, in the 1930s. Photograph from the H. Robins Hollyday Collection at the Talbot Historical Society, Easton, Maryland.

Figure 5.6. Live terrapins packed in barrels ready for shipment, with "seaweed" (or sea grass) used as a packing material. Photograph from the H. Robins Hollyday Collection at the Talbot Historical Society, Easton, Maryland.

By 1888, large females with eggs sold in Baltimore for up to $60 to $72 a dozen. Six years later, animals of 7 to 8 inches cost Baltimoreans about $100 a dozen. By 1904, prices were upward of $120 a dozen for "the largest and best quality" animals.

Because few people could identify real terrapin meat in a dish, the increased cost and scarcity of animals led to "terrapin" dishes being adulterated with meat of other turtles, including freshwater turtles.[18] Veal and chicken chunks might also be added. Pigeon eggs were sometimes substituted for terrapin eggs, and artificial terrapin eggs could be fabricated from hard-boiled chicken eggs.[19]

Artificial Propagation or Farming

In addition to harvesting the adults, people collected eggs from nests as food, thereby short-circuiting population recoveries. By the turn of the twentieth century there were fears that the populations had fallen too low—in 1886 an estimated 500,000 terrapins were harvested from the Bay alone, whereas 18 years later the yield along the Atlantic was not expected to be more than 12,000, with only 3,600 expected to be from the Bay.[20] The decline of Maryland's terrapin industry is illustrated by the fact that there were 57 terrapin dealers in 1904, 39 in 1916, and 7 in 1937.[21] Attention turned to farming to supplement the natural harvest.

Although terrapins penned on hard floors brought a lower price than wild animals when the latter were abundant, the increasing scarcity of wild animals undoubtedly forced acceptance of penned animals.[22] The damage to claws and shells caused by holding animals on hard floors could be avoided by providing sand. Some earlier entrepreneurs held terrapins in estuarine ponds, waiting for markets to improve. For example, Julius Ducatel wrote in 1837 about Somerset County residents "parking" terrapins in a square ditch excavated to allow for tidal movement and surrounded by 3- to 4-foot plank walls. Terrapins held in their thousands in these ponds were fed clams, crabs, and other inexpensive food until they were sold at higher prices in winter.

Subsequently, entrepreneurs went beyond holding captured animals for later sale. Some imaginative persons provided seminatural ponds with beaches protected from predators for holding captured terrapins and for rearing their offspring. This activity could be called farming. Charles Lewis on Hog Island in the Potomac River reportedly had thousands of animals in stock after five years of farming. In Crisfield, Maryland, a town built from the monetary rewards of harvesting oysters and blue crabs, A. B. Riggin & Co. and A. T. LaVallette & Co. farmed terrapins.[23]

In particular, Albert LaVallette moved from Philadelphia in 1887 to live on Hammock Point near Crisfield and rear terrapins in ponds he constructed on a 7-acre property.[24] He was initially very successful, buying terrapins from local watermen, holding them in the ponds where they were fed waste from local crab-picking plants, and then selling them to high-end restaurants in cities like Philadelphia and New York. He provided those restaurants with recipes for terrapin dishes and signed sole-supplier contracts with them. His business thrived. In 1890, he provided 28 barrels of terrapins worth about $4,700 for a banquet in New York's Delmonico's restaurant hosted by railroad baron Jay Gould. Terrapins nested in his ponds, which eventually held an estimated 15,000 animals of all sizes. LaVallette benefited from the increasing prices caused by falling supply, but eventually the demand for terrapin dishes waned and his business ended in the early 1900s. Plus, he ran away with the family's young governess.

Although farms attracted much attention from entrepreneurs and from tourists (figure 5.7), not all were successful.[25] For example, encouraged by a visit to a successful farm in Alabama, Robert Lowry started his own farm about six miles from Easton, Maryland, in 1885.[26] The site had a 1.5-acre pond connected by an inlet to

Figure 5.7. Catching terrapins by probing the marsh floor *(top)* and feeding them *(bottom)* on a Maryland farm. *Harper's Weekly,* February 25, 1888. Courtesy Calvert Marine Museum, Solomons, Maryland.

the Chesapeake Bay and into which a small stream emptied. He enclosed the pond and a small amount of dry land with a board fence. In May 1885 he added about 1,000 terrapins to the farm and fed them chopped blue crabs as well as the occasional fish. In his 1888 article he provided no information about the success of the farm. In 1904, a *New York Sun* article reported that a terrapin pond below Easton, Maryland, where a stream ran into the Bay had been provided with 1,000 animals, but after six months few animals could be found. Perhaps this was Lowry's farm. This article also reported that about 10,000 young animals were held in an enclosure near the mouth of the Potomac River around 1899, but only about 500 remained after two years. The fate of these terrapins is unclear, but terrapins are escape artists that can burrow below shallow fences or climb over them. In addition, rats, raccoons, crows, and gulls prey on terrapin eggs and hatchlings, wiping out a farm without protective netting over the beach nursery.[27]

In spite of limited successes with farming, interest persisted as harvests declined. Commissioner Hugh Smith wrote in 1895 that the US Fish Commission received many inquiries about the feasibility and methods of propagating terrapins because of the increasing scarcity of the animal in most Atlantic coastal states.[28] The commission was not then sponsoring propagation efforts, although it was aware of established farms in the Middle and South Atlantic states. Once laid in an appropriate beach-like habitat, terrapin eggs seemed very robust, needing no maintenance until they hatched. However, the commission knew that the main hindrances to terrapin farming included the animal's very slow growth rate and a lack of understanding by farmers of culture conditions that would support successful reproduction.

In succeeding years, terrapin abundances continued to decline until finally the Fish Commission (later the Bureau of Fisheries), in conjunction with the North Carolina Geologic and Economic Survey, began a study in 1902 in Beaufort, North Carolina, conducted by Dr. Robert Coker.[29] Eventually large tanks were built in Beaufort, and research on culturing terrapins from eggs continued until 1949.[30] However, with the declining interest in terrapins as a food, federal appropriations for terrapin research ceased in 1948. The Beaufort Laboratory released most of their broodstock, retaining a minimal number of animals until 1954, when the Laboratory finally stopped feeding those holdovers from the defunct program; in that year Maryland state biologists retrieved the few remaining adults.[31] I have found no record of what they did with the animals.

Managing the Resource

Concerns about managing the terrapin fishery while conserving populations grew over time. The first terrapin conservation law in Maryland was passed in 1878, setting a minimum under-shell length limit of 5 inches, declaring a closed season from April 1 to November 1, and outlawing possession, destruction, or disturbance of eggs.[32] Maryland continued to allow harvests over the next 129 years. Sometime after

1878, the Maryland Game and Fish Protective Association examined the state's terrapin laws, finding them "good, but they are irregular and they are not enforced."[33] The association recommended a $10 fine for terrapins possessed between April 1 and November 1 (this would have hampered farming, as it would not allow people to capture terrapins in warmer months and hold them in ponds in anticipation of winter sales). Other recommendations included a law limiting terrapin harvest in county waters to county residents. These recommendations seem to have been ignored.

Most changes in regulations after the association's report involved the open season, which varied from three to nine months in duration. In 1982, the minimum under-shell length was increased from 5 to 6 inches. Ultimately, the fishery declined to the point that the state banned the harvest of terrapins in 2007, although today it allows farming under a permit. Terrapins breathe air, so crab pots have to be fitted with a rectangular device or "terrapin excluder" to prevent terrapins from entering the submerged pot and subsequently drowning.[34]

American demand for terrapins as a food item declined greatly in the late 1920s. One suggestion for the diminished demand was that the onset of Prohibition removed access to Madeira or sherry, which enlivened the dish. Another was that the Great Depression inhibited high living. Turtle expert Archie Carr thought it more likely that the exaltation of terrapins as a gourmet food was a fad that had petered out.[35]

Now that harvests in the Bay are prohibited and crab pots have to have devices to keep terrapins from entering the pot, chances are good that populations will rebound, albeit slowly. Maybe the best baseline we can aim for is that "when one travels in a ship, their heads can be seen everywhere coming out of the water."

Uncontrolled Market Hunting of Waterfowl

A Mass Slaughter

～～～～～～～

The Indians had no other way of taking their water or land fowl, but by the help of bows and arrows. Yet so great was their plenty, that with this weapon only they killed what numbers they pleased. And when the water fowl kept far from Shore (as in warmer weather they sometimes did) they took their canoes, and paddled after them.

Beverley (1722b)

IN NORTH AMERICA, most migrating birds use one of four major north-south flyways. The most easterly is the Atlantic flyway, and the Chesapeake Bay lies along that route. The Bay is a major destination for many species of waterfowl (swans, geese, and ducks) as they migrate south in the autumn. So, although smaller numbers of waterfowl live on or around the Bay in summer, numbers increase greatly in autumn when migrants arrive after breeding in northern regions.[1]

A major plant food for some species of ducks is submersed wild celery *Vallisneria americana* that grows on the extensive flats near the mouth of the Susquehanna and elsewhere on the Bay shore and tributaries.[2] The flats are about 3 to 4 feet deep, and they covered an estimated 16 square miles in the late nineteenth century.[3] The roots of wild celery that grow in this region were, until the plant's recent decline, an especially important food item for the tasty canvasback duck *Aythya valisineria* (figure 6.1) whose scientific name reflects its appetite for the plant.

Many species of ducks and geese are hunted today, and management agencies try to estimate population numbers in order to set bag limits. However, when Europeans arrived in the Bay region and up until the late 1800s, there were few or no attempts to count numbers. It was not until the late 1800s and early 1900s that clear evidence of overharvesting stimulated conservation and management efforts.[4] Consequently, we must rely on anecdotal rather than quantitative accounts to describe the early popula-

Figure 6.1. Canvasback drake. US Fish and Wildlife Service photo by Lee Karney.
https://digitalmedia.fws.gov/cdm/ref/collection/natdiglib/id/6798.

tions. Nevertheless, earlier writers like Edward Forbush in 1912 collected these anecdotes to prepare persuasive accounts of the enormous abundances of these and other birds (not just waterfowl) seen by early explorers throughout North America.

These anecdotes mention extensive populations of wildfowl when early colonists were exploring the Chesapeake Bay watershed. For example, Captain John Smith wrote in 1608 that his exploration of the Chickahominy River in Virginia revealed "More plentie of swannes, cranes, geese, duckes, and mallards and diuers sorts of fowles none would desire." The seasonality of waterfowl numbers was remarked on by a number of early colonial writers through the 1600s, including Captain Smith in 1612:

> In Winter there are great plenty of Swans, Craynes, gray and white with blacke wings [whooping cranes], Herons, Geese, Brants, Ducke, Wigeon, Dotterell [small plover], Oxeies [small sandpiper], Parrats and Pigeons. Of all those sorts great abundance, and some other strange kinds to us unknowne by name. But in sommer not any or a very few to be seene.

Reverend Alexander Whitaker, writing in 1624, also described the winter populations:

> In Winter . . . the rivers and creekes bee over spread everywhere with water-foule of the greatest and least sort, as Swans, flocks of Geese & Brants, Duck and Mallard,

Sheldrakes [canvasbacks], Dyvers [loons], &c. besides many other kinds of rare and delectable birds, whose names and natures I cannot yet recite.

Forty years later, the seasonal profusion continued to amaze, as George Alsop wrote in his 1666 work commending colonial Maryland to potential settlers:

The Swans, the Geese and Ducks (with other Water-Fowl) derogate in this point of setled residence; for they arrive in millionous multitudes in Mary-Land about the middle of September, and take their winged farewell about the midst of March.

Even after decades of European colonization, winter waterfowl continued to be plentiful, as reported by colonial surgeon Thomas Glover in 1676:

On the Bay and Rivers feed so many wild fowl as in winter time they do in some places cover the water for two miles; the chief of which are wild Swans and Geese, Cormorants, Brants, Shieldfowl [Shelduck, or Canvasbacks], Duck and Mallard, Teal, Wigeons, with many others.

Ten years later in 1686, French Huguenot refugee Durand of Dauphine remarked:

One sees on the shores of the seas and on the banks of the rivers wild geese in troupes of more than 4,000 at a time . . . Ducks appear in flocks of more than 10,000.

As a final example, historian Robert Beverley wrote of the ease of killing great numbers (1722b):

As in summer the rivers and creeks are filled with fish, so in winter they are in many places covered with fowl. There are such a multitude of swans, geese, brants, sheldrakes, ducks of several sorts, mallard, teal, blewings [blue wing teal], and many other kinds of water fowl, that the plenty of them is incredible. I am but a small sportsman, yet with a fowling piece I have killed above twenty of them at a shot.

Around the same time as Thomas Glover's 1676 account of 2-mile-wide rafts of wild fowl, Jasper Danckaerts and a colleague explored Maryland east of the Chesapeake Bay, seeking land for a Dutch religious sect to purchase. In his journal of 1679–1680, Danckaerts mentions being unable to sleep because of waterfowl "screeching" during the night of December 3, 1679, on the Bohemia River, a tributary of the Elk River in the upper Chesapeake Bay:

We were directed to a place to sleep, but the screeching of the wild geese and other wild fowl in the creek before the door, prevented us from having a good sleep.

A week later, on December 10, he reported that near the Sassafras River:

I have nowhere seen so many ducks together as were in the creek in front of this house. The water was so black with them that it seemed when you looked from the land below upon the water, as if it were a mass of filth or turf, and when they flew up there was a rushing and vibration of the air like a great storm coming through the trees, and even like the rumbling of distant thunder, while the sky over the whole creek was filled with them like a cloud . . . they [local residents] are accustomed to shoot from six to twelve and even eighteen and more at one shot.

He continued to be astounded by waterfowl numbers, reporting on December 12 in the same region:

I must not forget to mention the great number of wild geese we saw here on the river. They rose not in flocks of ten or twelve, or twenty or thirty, but continuously, wherever we pushed our way; and as they made room for us, there was such an incessant clattering made with their wings upon the water where they rose, and such a noise of those flying higher up, that it was as if we were all the time surrounded by a whirlwind or a storm. This proceeded not only from geese, but from ducks and other water fowl; and it is not peculiar to this place alone, but it occurred on all the creeks and rivers we crossed, though they were most numerous in the morning and evening when they are most easily shot.

The thunderous sounds of massed beating wings persisted into the nineteenth century. Writing in 1833, unknown author S. H. reported that wintering ducks arriving at Havre de Grace covered acres of water, rising when disturbed "in flocks that darken the air; and the noise of their wings can be heard five miles or more on the water, resembling distant thunder."

A sense of how a large flock of ducks would appear taking to the air can be obtained from figure 6.2 even though that flock is small compared with what was reported

Figure 6.2. Pintails at Barren Island, Maryland, in 1930. Courtesy Chesapeake Bay Maritime Museum, Dr. Harry Walsh collection.

two or three centuries ago. Imagine scaring such flocks and hearing the thunder of their wings for mile after mile as you tramp along the water's edge. A baseline has changed greatly.

Commercial Hunting Grew Out of Control

Native Americans and early colonists hunted wildfowl for subsistence; later, wealthy individuals hunted for sport, generally eating what they shot.[5] Sport hunting, while not wasting the meat, could kill great numbers of wildfowl. In his 1833 report, anonymous author S. H. wrote that he saw 42 canvasbacks and redheads killed and 15 wounded by the simultaneous discharges of three shotguns into a flock of ducks swimming near a blind, with a party of three gunners on another occasion killing 76 ducks with three shots apiece and then 95 ducks with four shots apiece. Decades later, Mackay Laffan described an 1877 sport hunting trip on the Bush River where his party of three shooters killed 96 ducks between sunrise and nine in the morning, with one six-barrel blast (three hunters with double-barreled guns) killing 20 canvasbacks and redheads at once.

But it took more than recreational hunting pressure to diminish waterfowl populations. Thus, while Dr. James Sharpless explained in 1830 that the Bay was renowned as the greatest resort of waterfowl in the United States and described "innumerable Ducks, feeding in beds of thousands, or filling the air with their careering, with the beautiful swans resting near the shore, like banks of driven snow," he noted that he had heard from a number of residents that abundances had decreased by half in the previous 15 years—the baseline had shifted in a short time. He proposed three reasons for the decline. One was that disturbing the waterfowl on their feeding and roosting sites scared them from the region. A second faulted the overuse of underwater gill nets that entangled and drowned diving ducks (this practice had stopped by 1851, perhaps because drowned ducks became waterlogged and insipid).[6] A third reason involved increased commercial or pot hunting, including the use of swivel guns and sink boats (described below).

Commercial hunting began as communities expanded inland and people living away from the coast still wanted to dine on Bay ducks. Pot hunters provided waterfowl to satisfy the market. In 1846, newspaper editor William Porter described a tactic of commercial hunters that was frowned upon by recreational hunters:

> But there is a class of men, poachers that shoot for the market, who make the greatest havoc with this game. They silently in the night-time paddle or scull small boats into the very midst of large flocks or beds of ducks, while they are feeding, and with a tremendous piece mounted on a swivel in the bow, slaughter immense numbers, often killing eighty or a hundred at a shot. This mode of destroying them is restricted by legislative acts, under severe penalties, but the difficulty of capturing or convicting these poachers is such, that most of them escape the penalties of the

law, and pursue their unhallowed avocation, notwithstanding the greatest efforts to apprehend them; and their only punishment is the repeated anathemas and just indignation of all true sportsmen.

Doctor Elisha Lewis wrote in 1851 and in 1855 that it was common for a hunter to kill 100 ducks in a day, relating tales of skilled "ducker" W. W. Levy who had bagged 187 ducks in a day and as many as 7,000 canvasbacks in 1846–1847. Lewis also reported that a Mr. Boyd killed 700 ducks in a single March, when ducks were scarcer (presumably because most individuals had flown north to their breeding grounds). Philadelphia physician Milnor Klapp wrote in 1853 that the same Mr. Boyd killed 271 canvasbacks and redheads in the spring of 1850 (presumably in one day) and 163 canvasbacks on November 10, 1852. That late autumn day in November was the same one during which a Mr. Holly killed 119 canvasbacks, which became part of the several thousand ducks said to have been brought into Havre de Grace to market that day.

Some entrepreneurs maintained a squad of gunners. For example, in 1867, Thomas De Voe referred to an article in the *Norfolk Herald* of January 8, 1857, reporting that Edward Burroughs, Esq., who owned a farm at Princess Anne on the shore of Back Bay in southwestern Virginia (now a national wildlife refuge), had hired 20 men during the autumn 1856 ducking season. By December 20, these men had used 23 kegs of gunpowder, plus shot, to kill "all the varieties of the duck species known in our latitude, such as canvas-back, red-heads, mallard, black ducks, sprigtails [pintails], bull-necks [ruddy ducks], bald-faces (or widgeons), shovellers, etc., to which may be added a good proportion of wild geese." These fowl were packed in barrels and shipped from Norfolk to New York by steamer. Usually from 15 to 25 barrels were shipped weekly, although 31 barrels were shipped in one productive week of 1856.

Figure 6.3. Duck-shooting—opening of the season. Drawn by C. G. Bush. *Scribner's Monthly* March 1872, page 640.

The slaughter continued. In his 2003 account of waterfowling in the Bay, John Sullivan quoted a Baltimore newspaper report that 10,000 ducks were shot on November 1, 1880, near Havre de Grace. An estimated 3,000 to 5,000 were shot on opening day in the same region in 1887. The opening day carnage in the early 1870s was lampooned in a cartoon of the time (figure 6.3).

Waterfowl Abundances Declined

This carnage could not continue without diminishing wildfowl populations. Indeed, anecdotes presented in an 1891 *New York Times* article lamented the declines in the numbers of canvasbacks in the Chesapeake region. Whereas at one time 7,000 canvasbacks had been killed on opening day in the upper Chesapeake, according to this account opening day kills declined to 5,000, then 2,000, and finally to just a few killed in 1891.

Noted hunter George Bird Grinnell wrote in 1901 that in earlier years (he did not say when), redhead ducks could be seen in Maryland's Eastern and Hogg Bays in February and March floating in miles-long rafts estimated to contain up to 50,000 individuals. After stating that similar abundances were once seen at Poplar Island Narrows and in the Choptank River, he reported being told that at a location called "Lou's Point" it had been common for an ox-cart to carry off the ducks shot "on favorable days" by the assembled hunters. When he wrote, such numbers were becoming a memory because of increased hunting pressure, especially by commercial hunters.

An example of a much smaller day's harvest near Havre de Grace around 1920 is displayed in figure 6.4. This harvest, not possible today because of bag limits, occurred long after the decline in abundances had set in.

Lower harvests led to higher prices for ducks on the market. For example, Commissioner Hunter Davidson of Maryland's Fishery Force reported in 1872 that canvasbacks sold for $1 and redheads for 62 cents, modest prices compared to what

Figure 6.4. One day's duck harvest at Havre de Grace, Maryland, around 1920. Courtesy Chesapeake Bay Maritime Museum, Dr. Harry Walsh collection.

was to come.[7] By 1877 some Baltimore City entrepreneurs began hanging ducks by their feet around the rim of a circular basket, with ice placed into the basket and the whole placed in a freezer.[8] In this way the marketers waited until they could sell the ducks for a higher price; in 1888 that was $8 a pair for canvasbacks and $4.50 a pair for redheads (the price difference leading to redheads often being surreptitiously served in place of canvasbacks).[9] Market demand was enhanced by the demand from England for canvasbacks for Christmas feasts. Indeed, raconteur George Hopper wrote in 1916 that Queen Victoria and, later, King Edward had standing orders for annual deliveries of canvasbacks and redheads from the Bay.

Commercial hunting persisted for decades. Night shooting was practiced by poachers near Havre de Grace using skiffs bearing "punt guns" (figure 6.5) or "night guns" mounted on a swivel on the bow.[10] One punt gun weighed 160 pounds and had a 1⅝-inch bore. These guns could fire from 1.5 to 2 pounds of shot, killing 75 to 100 ducks at a time.

Figure 6.5. Dr. Harry Walsh with two punt guns used by market gunners. Courtesy Chesapeake Bay Maritime Museum, Dr. Harry Walsh collection.

Figure 6.6. Battery gun used by market gunners. Courtesy Chesapeake Bay Maritime Museum, Dr. Harry Walsh collection.

A battery gun (figure 6.6) involved an array of as many as 10 barrels covering an arc and primed to shoot simultaneously. Commissioner Davidson knew of a "cart load" of waterfowl killed with one use of such an array in 1872.

Sink boxes (also known as a sink boat, sneak boat, surface boat, or coffin-boat; figure 6.7) were used in daytime in concert with numerous decoys.[11] A shooter was accompanied by a colleague in a sailboat or rowboat waiting some distance away to collect the dead ducks and stand by in case the sink box, with minimal freeboard, sank.

David Fitzgerald describes a sink box, available for rent with a guide, license, guns, decoys, and a boat for $100 in 1894:

> Your first glimpse of a sink-box will not inspire you with confidence . . . Imagine a
> board platform, ten feet long by six feet wide, with a coffin let into the center until
> it is flush with the deck, and you will have a very correct notion of a sink-box.
> Around the edge of the platform there is a framework, over which canvas is
> stretched, to minimize the wash of the waves over the floating structure. Decoys
> are placed in an artistic arrangement, known to your guide, on the platform and on
> the canvas outworks, as well as grouped on the water, about twenty yards in front
> of the box. The sink-box is simply an appliance, which, by placing the gunner below

Figure 6.7. Postcard of duck hunter Jesse Poplar in a sink box surrounded by decoys, late 1800s. Courtesy Chesapeake Bay Maritime Museum, Dr. Harry Walsh collection.

the surface of the water, prevents the ducks from seeing him until the last moment. Lying flat on your back in the box, you are very effectually hidden from a low flying bird until it arrives in your immediate vicinity.

Sink boxes were thoroughly described by Sullivan in 2003. He noted: "The sink box was designed for one purpose—to assist hunters in killing ducks—and its success was unprecedented."

In 1916, raconteur Hopper enthusiastically described the work of Captain Bill Dobson, who was able to call in ducks while lying in a sink box. Dobson used three muzzle-loaders to slay the incoming ducks, bagging almost 500 canvasbacks and redheads in a single day. Hopper reported that more than 175 sink boxes were licensed for use on the Susquehanna Flats in 1915.

Despite the great numbers of waterfowl being killed while wintering in the Bay, some authors disputed the concern that populations might become extinct. In 1851, Dr. Elisha Lewis proposed that, since ducks bred in large numbers in remote regions, the population would continue to be replenished, a view that depended both on the ducks being safe from hunting while flying to and from their breeding grounds and remaining undisturbed on those breeding grounds. However, by 1889, ducks in the upper Chesapeake Bay, especially canvasbacks, were "unprecedentedly scarce" and the season a "failure."[12] One reason was that the wild celery had been smothered by sediment from spring floods the previous year. However, another was said to be the excessive hunting of ducks by members of "ducking clubs" that owned hunting sites along the Gunpowder, Bush, Middle, and Back Rivers in the upper Bay. Pot hunting by night and the use of hundreds of sink boxes by day must also have contributed to the decline.

Eventually, laws were passed in Maryland to limit hunting season length, types of guns allowed, and numbers of ducks possessed.[13] In 1887, the season was limited to November 1 through April 1, with shooting on Monday, Wednesday, and Friday only and not before five in the morning.[14] Starting too early could result in a $25 fine. Baltimore Police Marshal Jacob Frey wrote in 1893 that

> The game laws which protect the ducks are very stringent. The Maryland State law prohibits the shooting of ducks in flocks upon their roosting or feeding ground, or elsewhere, from a boat of any kind, an exception being made in favor of citizens of the counties bordering on the waters where the ducks are. These can shoot, while the ducks are flying, from any boat except a sneak boat or sink boat. Nor can they shoot from a blind or artificial point more than a hundred yards from shore. No one, according to this State law, can shoot over Chesapeake waters with any kind of a gun except one which can be conveniently carried upon the shoulder.

In spite of such laws, overharvesting continued. William Hornaday, director of the New York Zoological Society, one of whose goals was "The Preservation of our Native Animals," developed a "brief but pointed inquiry into conditions affecting bird life as they exist to-day throughout the United States." This survey was sent to a number of observers in each state and territory in 1897 and asked questions about declines in bird numbers over the previous 15 years and the causes of any decline. All of the 28 complete responses revealed a decrease in bird life in general, not just waterfowl (small birds like shore birds, bobolinks, and thrushes were also shot for food or sport in the eighteenth and nineteenth centuries). The declines ranged from 10% to 77%, with an overall average of 46%.[15] Reports from Maryland, Virginia, and Delaware were incomplete, but declines in surrounding regions included 33% in the District of Columbia, 37% in New Jersey, and 51% in Pennsylvania, so undoubtedly the Bay waterfowl also declined over time.

The respondents blamed the declines in bird life on a number of factors, but the most cited causes included sportsmen and "so-called sportsmen"; other shooters, including boys; pot-hunters; plume hunters (bird feathers on women's hats were the rage); and egg collectors, chiefly small boys. The report recommended conservation actions, including banning unlicensed egg collecting and commercial traffic in wild birds and game (save for fur-bearing mammals), as well as banning spring shooting. The report also recommended establishing hunting licenses and setting bag limits.

Not everyone agreed completely with Hornaday's report. In particular, a comment in 1898 by anonymous writer W. S. in an ornithological journal remarked on the difficulty in estimating changes in numbers over a 15-year period based on memory rather than hard data. Nevertheless, W. S. agreed that abundances of game, plume birds, and some small birds were declining and that Hornaday's recommendations about prohibiting both egg collecting and the sale of game, birds, and eggs should become law. In 1901, George Grinnell also recommended that spring shooting stop and that daily and seasonal bag limits be required. In addition, efforts were made to educate children about the economic value and beauty of birds and to encourage ways to attract birds with bird baths, feeding stations, and vegetation.[16]

Maryland legislators tried periodically to end night shooting, to regulate or ban the use of sink boxes and swivel guns, and to ban commercial hunting, but commercial hunters were politically influential for a time.[17] There was a small victory when laws passed in 1882 enabled a gang of poachers from Havre de Grace to be broken up and their guns destroyed. Eventually, conservation concerns carried the day nationwide.[18] The United States and Canada jointly passed the Migratory Bird Treaty Convention in 1916 and the Migratory Bird Treaty Act of 1918.[19] The latter act banned hunting with guns larger than 10-gauge and extended protection to bird parts, including eggs, nests, and feathers.[20] Over time, North American bird populations have begun to recover, and numerous people and organizations are working to return numbers to higher baselines. However, it will take a long time, if ever, for George Alsop's "millionous multitudes" to descend on the Bay in winter.

Sturgeon

A Prehistoric High Jumper Fell from Memory

On the white sand of the bottom
Lay the monster Mishe-Nahma,
Lay the sturgeon, King of Fishes;
Through his gills he breathed the water,
With his fins he fanned and winnowed,
With his tail he swept the sand-floor.
There he lay in all his armor;
On each side a shield to guard him,
Plates of bone upon his forehead,
Down his sides and back and shoulders
Plates of bone with spines projecting!

From "The Song of Hiawatha" by Longfellow (1855)

STURGEON OF VARIOUS SPECIES have been important in cultures around the world, with their huge adult size and armor of bony plates covering their body stimulating artists to write songs and stories about them. In North America, Henry Wadsworth Longfellow composed his classic poem "The Song of Hiawatha," and some Indian legends and dances centered on sturgeon.[1]

More prosaically, sturgeon bodies provided a variety of commercial products. The most profitable uses involved the beast's meat and caviar (their preparation is described below). The head, skin, and spine were rendered to produce fine oil used in tanneries. The swim bladder provided the makings of isinglass, a gelatinous substance for clarifying jellies, glues, wines, and beer. Some entrepreneurs transformed spinal cords into a strong twine. Any residue left over from these activities might be used as fertilizer.[2]

Two sturgeon species live along the western Atlantic coast—the larger Atlantic sturgeon *Acipenser oxyrinchus* (figure 7.1) and the smaller shortnose sturgeon *Acipenser*

Figure 7.1. Atlantic sturgeon. US Fish and Wildlife Service illustration by Duane Raver.

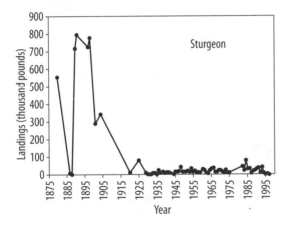

Figure 7.2. Landings of sturgeon in Chesapeake Bay from 1880 to 1997. Data from Fisheries Statistics Division 1990.

breviroster. Numbers within both species have declined to the point that they are listed as endangered. The Atlantic sturgeon, the subject of this chapter, was once the focus of coastal fisheries, including those in the Chesapeake Bay and especially those in Delaware Bay and River, with the smaller shortnose sturgeon an incidental catch.[3]

Overfishing, degradation of spawning and nursery habitat, incidental capture and killing of juvenile sturgeon in shad and river herring nets, and obstructions like dams blocking access to upstream spawning grounds all contributed to the decline in the numbers of sturgeon in the mid-Atlantic region.[4] The fishery in the Bay was smaller than the Delaware fishery; however, its steep decline (figure 7.2) adds to the story of changing baselines.

Atlantic sturgeon are anadromous, like shad and river herring. Adults spend much of their time in coastal salt waters, entering the Chesapeake Bay in spring and early

Figure 7.3. Three men retrieve a net containing a struggling sturgeon as large as the boat. WPA painting "A View of Hyde Park Landing" by Olin Dows (1941) in the series Professions and Industries, Hyde Park (New York) Post Office. www.hydeparklanding.com/history.html.

summer to swim upriver to spawn. Adults can grow as large as 14 feet and, if trapped in nets, can tear holes in the net as they struggle to free themselves (figure 7.3). This net damage led to many fishers killing and discarding captured fish before the flesh had commercial value.[5]

Early settlers in the Chesapeake region remarked on the presence of numerous large sturgeon, which resembled the fish they were familiar with back home but whose numbers were depleted. Colonial diarist William Byrd reported in 1728 that Indians caught sturgeon either with a net on a long stick or by placing a noose over the tail of an unsuspecting specimen and hanging on tightly. Robert Beverley's more complete description of the latter technique (1722a) is worth appreciating:

> The Indian way of catching sturgeon, when they came into the narrow part of the rivers, was by a man's clapping a noose over their tails, and by keeping fast his hold. Thus a fish finding itself entangled would flounce, and often pull the man under water, and then that man was counted a cockarouse, or brave fellow, that would not let go; till with swimming, wading, and diving, he had tired the sturgeon, and brought it ashore.

The Jamestown colonists reported that sturgeon were seasonally plentiful in the James River, as these reports illustrate:

> For foure monethes of the yeere, February, March, Aprill and May, there are plentie of Sturgeons. (Hariot 1590)
>
> Sturgeon, great store, commonlie in Maie if the yeare be forward. I have beene at the taking of some before Algernoone fort, and in Southampton river, in the middst of March, and they remaine with us June, July, and August, and in that plenty. (Strachey 1610)

In sommer no place affordeth more plentie of Sturgeon, nor in winter more abundance of foule, especially in the time of frost. I tooke once 52 Sturgeons at a draught, at another 68. From the later end of May till the end of June are taken few, but yong Sturgeons of two foot, or a yard long [the adults would be in fresh water spawning, with the juveniles in the estuary]. From thence till the midst of September, them of two or three yards long and few others [the adults were returning to the sea]. And in 4 or 5 houres with one Net were ordinarily taken 7 or 8; often more, seldome less. (Smith 1624)

We had more Sturgeon, than could be devoured by Dog and Man, of which the industrious by drying and pounding, mingled with Caviare, Sorrell and other wholesome herbes would make bread and good meate. (Smith 1624)

They were also abundant elsewhere in the Bay:

It [the site of Washington DC] aboundeth in all manner of fish. The Indians in one night commonly will catch thirty sturgeon in a place where the river is not 12 fathom broad. (Fleet 1632)

The accounts of abundant sturgeon continued into the 1700s, such as the report of the Reverend Burnaby quoted on page 6 in chapter 1. A particular behavior of sturgeon that entertained boat passengers in east coast estuaries is that even the largest sturgeon is capable of prodigious leaps out of the water. Swedish naturalist Peter Kalm visited North America and remembers that on May 31, 1749:

About noon I left Philadelphia, and went on board a small yacht which sails continually up and down upon the river Delaware, between Trenton and Philadelphia . . . Sturgeons leaped often a fathom into the air. We saw them continuing this exercise all day, till we came to Trenton.

This behavior, while undoubtedly exciting to travelers, could sometimes be dangerous. Historian Jonathan Elliot reported in 1830 that a sturgeon took to the air as the Georgetown ferry from Washington, DC, to Virginia was passing. The sturgeon fell onto the leg of a passenger and broke it, which subsequently led to the unhappy man's death. There are many other anecdotes about high-jumping sturgeon in the Delaware and Hudson Rivers, with some injuring passengers on boats.[6]

As Captain Smith noted, sturgeon were eaten by the first colonists. However, after terrestrial farming and animal husbandry became established in the Bay's watershed, sturgeon were eaten less. There is one report by John Burk in 1805 of an early Virginian fishery that had failed:

A sturgeon fishery is spoken of as existing at this time [1626] in Virginia, the expenses of which, at this date, were stated to be 1700£. We are not informed whether this establishment was set on foot by the government, or by private adventurers. The experiment however did not meet the expectations of the projectors, and it appears to have been about this time abandoned.

Although this early development of a fishery in the Chesapeake Bay was short-lived, eventually a modest fishery revived in the early 1800s, if not before. As reported separately by Jonathan Elliot and Joseph Martin, sturgeon in the early 1830s were a food item in the James and Potomac River regions, attaining an average weight of from 40 to 150 pounds.[7] They were captured in large-meshed floating nets or with a special iron hook built for snagging their body.[8] The latter device was 30 inches long, including a 24-inch stem, and had a barb on the inside of its curve. A 3- or 4-pound weight capable of sliding back or forth was attached to the hook to weigh it down. The device was paid out to near the estuary bottom and was either moved back and forth by rowing the boat slowly or just allowed to hang straight down with the boat at anchor. The fisher held the line so that when it bumped against a sturgeon or when a sturgeon blundered onto the hook, the fisher felt the motion and tugged the line upward rapidly. The combination of the hook's curvature and the position of the weight situated the hook directly under the fish, where it penetrated the thin, unprotected belly. A large, hooked sturgeon was capable of towing a small boat for some distance until the fish became exhausted and could be captured. This illustrates how strong and persistent an Indian cockarouse had to be to land a large snared sturgeon.

This mode of hooking sturgeon and the fact that the flesh sold for three cents a pound suggests that sturgeon were very abundant at the time.[9] Or, the low cost of the meat might mean that the fish was not appreciated as a food. Indeed, John Cobb in 1900 and Walter Tower in 1908 reported that in most regions until the 1800s sturgeon flesh was fed predominantly to servants and slaves, it being considered unsuitable for others. In 1899, Charles Stevenson's extensive review of food preservation confirmed that in the early nineteenth century there continued to be limited use of sturgeon as food.[10] That situation changed gradually, beginning on a small scale in New York City around 1857 and then continuing in Philadelphia, as entrepreneurs smoked the meat and collected the roe to be sold to Europeans (especially Germans and Russians) as caviar.[11]

Markets expanded over time, with sturgeon meat prepared as steaks, like halibut, with the skin removed.[12] Young fish were preferred because the yellow fat of larger fish had a stronger flavor. The meat had a beefy appearance, leading to Virginians calling it "James River Bacon" or "Charles City Bacon."[13] The flesh could also be roasted, baked, or fried, usually after scalding it in hot water to dissolve any fat.[14] Some flesh was salted, pickled, or smoked, with fish to be smoked often held pickled in barrels until needed. By the time Stevenson assembled his comprehensive report in 1899, most sturgeon in the United States were smoked, with German immigrant populations in Buffalo, Sandusky, Milwaukee, Chicago, Philadelphia, and New York City representing the largest markets.

Initially, there was no market for the roe, which was used to feed pigs or as fish bait.[15] Sir Augustus John Foster, a British representative to the United States in the early 1800s, recorded his own attempt to make caviar an acceptable dish in Washington, DC.[16] He wrote:

Plenty of sturgeon are caught at the little falls of the Potomac a short distance above George Town where the river becomes narrow and the scenery is very romantic; such abundance was there indeed of this fish that I determined to try if the roe might not be cured so as to afford Caviar and my maitre d'hotel having nothing to do in the summer, I gave him a receipt out of Chambers Dictionary for the purpose which he so successfully followed that I had some excellent Caviar for the following winter but on its being served to the members of Congress, the precaution of telling them to taste a little first not having been observed they took such quantities thinking it was black raspberry jam that the stock was soon exhausted and very few of them liked it but spit it out very unceremoniously as a thing excessively nasty. Nevertheless it had met the approbation of some of the gentlemen of the Russian legation and I trust that the manufacture of it being thus introduced into the country it may by degrees become an object of consumption and even of exportation.

In 1908, Walter Tower documented the history of sturgeon fisheries nationwide. He wrote that it was not until after 1850 that regular fishing began, probably in the Delaware River region, when the fish sold for very little money (12½ to 30 cents each). He cited a report in the *Fishing Gazette* that so many fish were harvested from the Potomac in the 1880s (probably as by-catch in shad seines) that there was no market for them, so they were piled on the riverbank and carted away by farmers to be used as fertilizer. Demand for sturgeon meat and roe in the United States gradually increased so that, over time, the price of sturgeon meat rose from less than 1 cent per pound in 1881 to 12½ cents per pound in 1896. From 1882 to 1884 a female fish could sell for $2; by 1889 the price had risen to $30 to $35. Nationally, a keg (130 to 135 pounds) of caviar sold for $9 to $12 in 1885, $20 in 1890, $40 in 1894, and over $100 by 1900.

The major mid-Atlantic fishery in the 1800s was in the Delaware Bay and River region. In 1884, sturgeon swam into the Delaware River in late May in such numbers that drift gill nets measuring one-quarter of the usual length of about 1,200 feet had to be used because larger nets would catch so many fish as to be unmanageable; the average catch was 25 to 30 fish in one placement of a net.[17] The nets had a stretch mesh originally measuring 16 inches; mesh size declined to 13 inches over a decade to capture smaller sturgeon as larger fish became scarcer. Many fishers lived and prepared their harvest on anchored scows during the spring fishing season, which lasted about two weeks.[18]

In 1880 the industry nationwide yielded about 6,000 tons of product, with the Delaware Bay region contributing about 10%. Harvests nationally continued about that level for the next 20 years, with the Delaware region's yield increasing to about 3,200 tons by 1888. However, there was a drop in the average numbers of sturgeon caught per gill net (a measure of fishing success) from 60 sturgeon in 1890 to less than half that amount in 1896.[19] In 1897, about 1,000 fishers were using more than

150 miles of gill nets in Delaware Bay. After 1897, the industry began collapsing to the point that a decade later there were few fish caught in the region or along the East Coast.

Declines in harvests were attributed to wasteful fishing, including the capture of juvenile fish of no commercial value, and harvesting during spawning season (because, of course, the eggs of pre-spawning females were the source of caviar).[20] In terms of environmental disruption, just as for shad and river herring, water pollution degraded spawning habitat, and dams prevented upstream migration to former spawning grounds.[21]

The rise and decline of the sturgeon fishery in the Delaware region mirrors the overfishing of oysters, shad, and river herring that occurred in the Chesapeake Bay. Presumably the sturgeon fishery in the Chesapeake Bay followed a similar boom and bust scenario. The available data are limited but show a decline over time. In the Potomac, 288,000 pounds of sturgeon were landed in 1880, 60,920 pounds in 1890, and 45,710 pounds in 1891.[22] In Maryland as a whole, 99,932 pounds were landed in 1890 and 72,445 pounds in 1891; in Virginia 817,670 pounds were landed in 1890 and 723,646 pounds in 1891.[23]

Magazine writer Charles Coleman described an artisanal fishery for sturgeon that still existed in the James River, Virginia, at the end of the nineteenth century.[24] He accompanied a team of two fishers who were part of an itinerant "company of fishermen" based on Jamestown Island. The company employed 18 boats, and the fishers worked from transient camps along the north shore of the James, April through summer, with a peak in May to early June. Coleman's companions used a gill net that was a half mile long and 30 feet deep and was paid out over the stern as the boat was rowed from the nearshore out to the center of the river. The net was suspended from about a hundred wooden floats and drifted after being fully deployed. Nets were set just before high and low water, with the average catch per day being about two sturgeon per boat. A strike from a sturgeon was indicated by one or more floats disappearing underwater. Captured fish ("cows" and "bucks") were towed to shore by a halter inserted through the mouth and out one gill, an action that apparently calmed the fish. Otherwise they would often thrash around, sometimes damaging a boat. On shore, cows had their roe removed, and the bucks were left to die before being transferred with the eviscerated cows and their roe to a steamboat for transport to market.

Hildebrand and Schroeder summarized the status of Atlantic sturgeon in the Chesapeake Bay around 1928, noting that the species had declined greatly in abundance and was now seldom found north of the Potomac River mouth. The authors had made inquiries around the Bay in 1921 and 1922 as to the species' status and were told that sturgeon had been scarce throughout the region for many years. People were aware that the baseline had shifted downward.

As the abundances of sturgeon dropped, there were attempts at restoring stocks artificially by hatching fertilized eggs. Such fish hatchery practices had been applied to shad, trout, and other fish species with varied success. For sturgeon, similar efforts were hindered by the difficulty of collecting ripe eggs and sperm simultaneously because of the scarcity of fish. Also, the recovery of the species was (and is) hampered by the fact that sturgeon are slow to become adults. In the Mid-Atlantic region, females mature at 10 to 20 years and males at 6 to 10 years.[25] In addition to the long time required before fish become adults, females spawn only once in a 2- to 6-year period. Given these realities, it will be some time before Atlantic sturgeon populations can be restored to a point that the fish cease to be endangered and their aerial high jumps once again become a common source of amusement.

Blue Crabs Hung On

~~~~~~~~~~~~~~

The sounds, rivers, creeks, and marshes in the vicinity of Crisfield may be said to teem with crabs. So great is the supply, in fact, that it seems almost inexhaustible.

Smith (1891a)

Between April and October, crabs are found in indescribable abundance, in all the bays and sounds, from the Chesapeake Bay southward. In many places they are so numerous that there is no market for them, and even in the Chesapeake Bay it is not unusual to see thousands dragged on to the shore and left to die or to make their way back into the water, by fishermen who have shaken them out of their seines and abandoned them. Further south the fishermen in the channels find their work so much obstructed by the crabs that they trample upon them, or crush them with clubs, to keep them from returning to the water to clog their nets again.

Brooks (1893)

HARD-SHELLED BLUE CRABS ARE feisty crustaceans, always ready for a fight (figure 8.1). They hold on tightly to a finger or a piece of bait, the latter action facilitating their capture by trotline (below). Soft crabs are molted hard crabs that are generally immobile or slow-moving while their shell hardens. In contrast to the steep declines in the fisheries described in earlier chapters, the fishery for hard crabs has increased (although harvests of soft crabs have declined; figure 8.2). Thus, this chapter will include information on the Bay's blue crab fishery into the twentieth century rather than stopping in the late nineteenth century.

Harvest statistics can be underestimates. As with oysters (see comment by the US Commissioner of Agriculture on page 45), crabbers who sell directly to the public (e.g., at a boat dock or from a roadside vehicle) may be overlooked. Thus, reported commercial harvests in Virginia have been estimated to be about one-third of the actual

Figure 8.1. Blue crab with its two front claws raised in an aggressive position. Courtesy of Skylar Hepner and Maryland Department of Natural Resources.

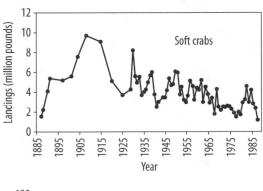

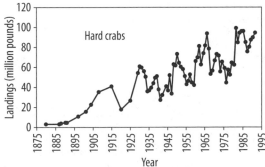

Figure 8.2. Landings of soft *(top)* and hard *(bottom)* crabs in Chesapeake Bay from 1887 to 1988 (soft crabs) and from 1880 to 1991 (hard crabs). Data from Fisheries Statistics Division 1990.

harvests. Recreational harvests are also difficult to estimate. Maryland's recreational harvest has been estimated to be 15% to 25% of the commercial harvest, with a lower percentage in Virginia.[1]

Blue crabs move throughout the whole of the Chesapeake Bay during their life cycle (figure 8.3). Adult males and females within the Bay become targets for commercial

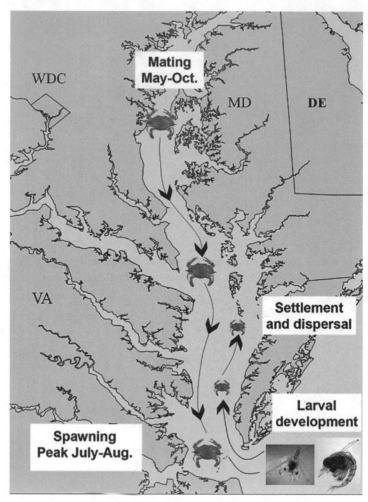

Figure 8.3. Diagram of the reproductive and developmental history of Chesapeake Bay blue crabs. © 2017 Smithsonian Environmental Research Center.

and recreational fishers who use the gear illustrated in figures A.8–A.10. During the summer and autumn, adults mate throughout the system, with fertilized females moving down-Bay to the higher salinity mouth where larvae hatch from the masses of eggs (colloquially called a sponge) that the females (now known as sponge crabs) carry under their abdomen. The larvae float and feed offshore, passing through a number of developmental stages before being carried back into the estuary by the high-salinity water that floods into the Bay at depth (see figure 2.4). Once in the Bay, the larvae metamorphose to become juveniles that disperse throughout the Bay as they grow and mature.

Brief History of the Fishery

In 1887, Robert Rathbun wrote the first comprehensive report of the US blue crab fishery, including details about fishing methods, harvests, and economic values for different regions from New England to the Gulf of Mexico. Here I focus on the Chesapeake Bay fishery, supplementing Rathbun's reports with information provided by Hugh Smith in 1891, William Brooks in 1893, Winthrop Roberts in 1905, and Edward Churchill in 1920.

In early anecdotal reports, some colonists mentioned blue crabs as food. Indian Chief Powhatan included crabs in a morning meal served to visiting Europeans in the early 1600s. A tutor on a plantation on the lower Potomac River wrote in his journal in 1774 about various meals that included crabs and fish. English traveler Richard Parkinson noted in the late 1700s that hard and soft crabs were eaten only near where they were caught.[2] Crabs were so abundant that they were used as bait for line fishing for finfish, as well as food for penned terrapins (see chapter 5).[3] As with abundant terrapins, in the early nineteenth century most people eating crabs were probably catching them themselves, with crabbing being a casual operation.[4] In Maryland, there were no laws about size limits or seasons, dip netting needed no license, and only Dorchester County required a license for scraping with a small dredge.[5] As the century progressed and market demand picked up nationally, increasing numbers of blue crabs were shipped to cities close to harvest sites, like New York, Philadelphia, Charleston, Savannah, Mobile, New Orleans, and Galveston.[6]

One reason for export originally being limited to cities near harvest sites was that it was difficult to keep delicate and perishable crab meat fresh or soft crabs alive out of water for any length of time in warm weather, which is when many crabs are harvested. Until the late 1800s, when there were improvements in mechanical refrigeration and ice-making technology, keeping food cold depended on natural ice imported from northern states.[7] Another factor hindering long-distance transport of crabs and their meat was the relative slowness of road and ship transport. Expanded railroad systems in the later 1800s (see chapter 1) allowed perishable seafood to move more quickly from coastal fishing towns to inland markets.[8]

By the late nineteenth century, the Chesapeake Bay blue crab fishery was becoming the largest of all the states, with the fishery for soft crabs being more remunerative than for hard crabs.[9] The most important shipping location in the United States became Crisfield, located near the junction of Pocomoke and Tangier Sounds in mid-Bay (see figure 4.4). The Eastern Shore Railroad linked Crisfield to Baltimore and Philadelphia in the late 1800s, and steamboat facilities were also available.[10] As a result, most of Maryland's harvest and a large component of Virginia's passed through the town. Nearby Deal Island was next in importance to Crisfield but only had steamboat facilities, not a rail line.[11]

In 1880, the combined soft and hard crab harvest from Maryland and Virginia was 3.8 million pounds, compared with 1.6 million from New York and 1.5 million from New Jersey. Chesapeake harvests grew to about 9.5 million pounds in 1890 and 21.5 million in 1901, eventually reaching 50 million in 1915. The importance of the Chesapeake fishery grew steadily in the twentieth century, attaining about 50% of the national market by midcentury.[12]

Soft Crab Fishery

The mass production of soft crabs seems to have begun in New Jersey around 1855, probably thanks to the proximity of New York City markets.[13] Maryland's soft crab industry began about 18 years later, with soft crabs first shipped outside Maryland (to Philadelphia by the new railroad) in 1873 or 1874 by a Crisfield firm.[14]

As Fish Commissioner Hugh Smith wrote in 1891, the Bay's soft crab industry depended on the grassy shallows of Pocomoke and Tangier Sounds. To capture soft crabs and peelers ready to shed in the sounds near Crisfield, hundreds of canoes (18 to 25 feet long and with one or two sails and a jib sail) crewed by two or three crabbers dragged two crab scrapes (or three if the winds were stronger) over the bottom. Scrapes were retrieved every few minutes to allow the soft crabs to be picked from the seagrass, mud, and oyster shell caught by the tow. Other crabbers pushed unrigged bateaux in shallow marshes, taking crabs in dip nets or small seines. These were time-consuming actions that yielded fewer soft crabs per day than the hard crab fishery, which used baited trotlines in open habitat where ravenous hard crabs abounded. Soft crabbers averaged 150 to 300 crabs per day.[15]

The industry became so profitable in Crisfield in the early years that 685 and 773 crabbers harvested soft crabs and peelers in 1887 and 1888, respectively, whereas only 12 crabbers harvested hard crabs in those years. By 1901, the numbers were 2,164 soft crabbers and 89 hard crabbers.[16] This disparity in fishing effort was related to the fact that the less common soft crabs were more in demand and thus more valuable in the aggregate than hard crabs. Prices to the soft crabber in 1887 and 1888 ranged between 18 and 30 cents per dozen, with shippers selling them to dealers in cities for 35 to 60 cents per dozen; selling hard crabs to shippers garnered about 6 or 7 cents per dozen.[17]

By 1884, soft crabs were packed in nested trays (usually 18 by 28 by 10 inches; figure 8.4) of thin pine boards with two or three trays that could be removed to reveal crabs in each tray.[18] Seagrass was placed on the bottom of a tray, and then three to eight dozen soft crabs (depending on crab and tray sizes) were packed tightly together nearly vertically and front end up (to help keep gill cavities moist) in each tray. The crabs were covered with more seagrass or moss and finely crushed ice. About 75% of crated crabs survived to market. Crabs packed carefully in Crisfield went to Baltimore for shipment as far north as Canada and west to Chicago. They could arrive alive in Canada after 60 to 72 hours.[19]

Over time, markets expanded greatly. In 1909, on some days up to 1,000 boxes holding 180 soft crabs each were shipped by express from Crisfield, with as many as

Figure 8.4. Live soft crabs packed in a wooden tray with "seaweed" (or sea grass) to retain moisture and ready to be covered with ice before shipment. Behind the trays are two tins of lump meat picked from hard crabs. Churchill 1920. Courtesy Calvert Marine Museum, Solomons, Maryland.

seven railroad cars of soft crabs being shipped daily by 1916.[20] Other regions also yielded soft crabs at lower volumes, with Deal Island shipping about 50,000 weekly, along with about 3,000 hard crabs. Efforts to add value to soft crabs began early; fried soft crabs were canned for foreign markets by 1877.[21]

To meet the demand for soft crabs, peelers were held in circular or square holding pounds built of wooden laths pushed into the estuary bottom and nailed closely together to keep crabs from escaping.[22] Peelers were fed and watched closely for molting. Pounds were difficult to manage, and numerous peelers were lost to cannibalism or died as a result of wide variations in temperature, salinity, or water quality (dead crabs were often fed to hogs or terrapins). Experienced producers began to equip their pounds with wooden floating boxes (floats) to house and protect peelers that were close to shedding (figure 8.5). Rathbun (1887) described crab shedding in such floats in northern New Jersey, a technology later adopted in Maryland.[23]

Firms that bought, shed, and shipped crabs in Crisfield might use up to 100 floats in a pound (25 was average) that could hold 300 to 400 crabs each.[24] Crabs in a similar shedding stage (e.g., early, late, molting) would be placed in the same float, examined two to four times daily, and moved as their shedding progressed. Soft crabs were

FLOATS FOR SOFT CRABS AT OXFORD, MARYLAND.

Figure 8.5. Soft crab floats being tended at Oxford, Maryland.
Stevenson 1899.

removed with minimal handling a few hours after they had molted (if removed too soon after molting they would die during shipping) and were packed for shipment.

Crab floats were relatively inexpensive to construct, maintain, and operate, but disadvantages outweighed advantages in a float operation. To begin, there had to be access to waterfront property where pounds could be located offshore while relatively sheltered from storms. Crabs in floats were confined to the upper few inches of water; at these depths, mortalities can occur from fluctuations in temperature, salinity, and dissolved oxygen caused by rapidly changing weather or heavy rainfall. In quiet, storm-protected areas, water circulation may be poor, so dissolved oxygen can decline, and water temperatures can rise to lethal limits. Crabs held in floats also were exposed to predation by fish, birds, and mammals. Essentially, there was little control over environmental factors.

Another drawback to a float operation was the physical discomfort associated with tending moored floats from a boat, with the operator bending over a gunwale and the edge of a float (figure 8.5). The desire for convenience led to the development of shore-based, flow-through systems—a move credited to Wellington Tawes of Crisfield around 1950.[25] Originally, shedding tanks were simple troughs or shallow tables placed on land with running water pumped from an adjacent brackish water supply and then returned overboard. Tanks were soon housed to provide shade and protection from rain, poachers, and predators and raised to waist level, where little bending is required. More stable temperature and salinity levels could be achieved by drawing water from greater depths offshore. For even greater control over the environment, recirculating water (closed) systems began appearing in the Chesapeake region in the early 1980s.[26]

Hard Crab Fishery

The disparity between prices for soft and hard crabs reversed slowly. In 1901, Maryland harvested 13 million soft crabs worth $203,000 and 29.5 million hard crabs worth $86,000; in 1904, these numbers were 17.2 million soft crabs worth $190,000 and 38 million hard crabs worth $169,000.[27]

As mentioned above, some hard crabs were harvested in Pocomoke and Tangier Sounds by a small number of crabbers, but the bulk of the hard crab fishery was located in the other tributaries of Virginia and Maryland (figure 4.4). Hard crabs were fished with cotton, manila, or grass rope trotlines of about 3/8-inch diameter, from 200 to 1,000 yards long, that were often protected by tar and were baited with eel heads, tripe, or "stinking meat."[28] In the winter in Virginia, torpid crabs buried in the Bay bottom were captured by dredging. Lastly, the wire crab pot was introduced into the Bay in the mid-twentieth century.

Live hard crabs were packed for shipment to processors mostly in barrels (figure 8.6) holding about 200 to 300 crabs (male crabs are larger than females) worth about 50 cents to $2 a barrel in 1901 and $1 to $1.50 in 1916.[29] As with soft crabs, efforts to add value to hard crab meat began early. Deviled crab became popular, to the extent it was said that the dish had made an enterprising Baltimore widow very wealthy.[30] Deviled crab was served in the upper carapace shell of a hard crab whose meat had been extracted (picked), and processors shipping crab meat in 1905 sent on average from 80 to 100 shells with each gallon of meat.

Hard crabs had their meat extracted in processing plants built near Eastern Shore crabbing grounds and in Baltimore (see tins of meat in figure 8.4). In these plants,

PACKING LIVE HARD CRABS IN BARRELS FOR SHIPMENT AT OXFORD, MARYLAND.

Figure 8.6. Adding "seaweed" (sea grass) to barrels of hard crabs packed for shipment, Oxford, Maryland. Stevenson 1899.

women and children picked crab meat for bulk shipment. Demand for crab meat started slowly, with about 395 gallons of meat picked from about 27,000 hard crabs worth $1.20 per gallon shipped from Crisfield in 1888. By 1910, a single processor near Crisfield shipped 1,500 gallons.[31]

As markets developed in the late 1800s, a hard crabber could make a good living in a relatively short day's work in season. Professor Brooks wrote in 1893 about the fishing effort of Virginia crabbers: "The abundance of the crabs in our waters is well illustrated by the fact that we were told, in 1884, by fishermen in the lower part of the Chesapeake Bay, that they were earning from $1.50 to $2.00 a day catching crabs to sell at one cent a dozen or ten cents a bushel; and these men seldom went to their work before sunrise or fished longer than till noon. In fact, most of them were home for the day at ten in the morning."

Brooks' data on the price per dozen translates to a morning's catch per crabber of 1,800 to 2,400 hard crabs. Such catches were obtained by trotlines deployed from sailboats, with the crabber (usually working alone) pulling the boat along the trotline by hand (gasoline engines became available around 1902).[32] By about 1950, catches by trotlines (now deployed from motor boats) had dropped to 600 to 1,200 crabs daily.[33] Brooks' report might seem exaggerated, but he was a trained scientist, so his harvest estimates should be reasonably reliable. With foresight, he noted that "The supply of crabs in our waters does not as yet show any signs of exhaustion, but the history of the lobster fisheries proves that the extension of the canning industry and the increased demand for crabs which this will produce must ultimately exhaust the supply. Measures for the preservation and protection of the crabs must some time be adopted."

This prediction that increased demand would jeopardize supply began to come true at the turn of the century, stimulating periodic examinations of the fishery and subsequent legislation to foster conservation.[34] For example, a decline in harvests in the early 1900s led the US Bureau of Fisheries to assign Edward Churchill the task of studying the Bay's blue crab fishery.[35] Churchill used data collected from 1907 to 1917 by two Virginia packing houses to determine catch per trotliner and catch per winter crab dredge (these are measures of fishing effort). Even though effort by trotliners rose because trotline lengths increased from 600 to 900 feet during the study period, the average daily harvest of hard crabs declined from about 4.6 barrels per crabber in 1907 to about 2 barrels in 1917. For the winter dredge fishery, catch declined from about 30 barrels per boat in 1906–1907 to about 9 in 1916–1917. As a conservation attempt, Virginia established a minimum size limit of 3½ inches for hard crabs (except for peelers preparing to shed) in 1912, raising the limit to 5 inches in 1916. Also in 1916, Maryland set limits of 5 inches for hard crabs and 3 inches for peelers and soft crabs.

The harvest declines sparked controversy between Maryland and Virginia about the harvesting of sponge crabs in Virginia waters. Most sponge crabs are found near the Bay mouth toward which the females have moved and where the eggs in the sponge eventually hatch. Various reports by Maryland's agencies remarked that the powerful crab packing industry in Hampton, Virginia, worked against legislation banning such harvests.[36] Eventually, in 1916 Virginia did ban the taking or possession of egg-bearing females in July and August, and Maryland banned possession of sponge crabs at any time of year.[37]

Further deterioration of the fishery led to another federally initiated survey of the Bay's crab fishery in 1924 and a report in 1925 by Elton Sette and Reginald Fiedler.[38] They found that the decline in average catch per trotline and crab scrape that Churchill had uncovered had continued. They recommended that capture and possession of sponge crabs be prohibited and that the fishing effort be reduced by 30%. They also recommended that statistical and biological data be collected continuously to monitor the effects of regulations on the fishery and that state fish commissions should be able to change regulations under government oversight. Virginia subsequently passed a law in 1926 outlawing possession of sponge crabs at any time.[39] Neither state attempted to reduce harvest pressure or to collect long-term data until Maryland began to collect harvest data in 1928.[40]

In 1942, Virginia established a 146-square-mile spawning sanctuary at the Bay mouth—the Virginia Blue Crab Sanctuary.[41] The sanctuary was first closed to recreational and commercial crabbing in July and August, with the closed period expanded from June 1 through September 15 in later years. The area of the sanctuary also gradually expanded over time. The area is intended to protect spawning females (it seems to have been successful in this), potentially boosting reproduction and expanding future crab populations.[42]

Contemporary Management Concerns

Contemporary fishery managers face many problems. For hard crabs, problems include those common to fisheries in crowded aquatic habitats that are heavily used for commercial and recreational activity, namely competition among user groups, overfishing, habitat degradation, and inadequate collection of assessment data, especially for the recreational fishery. "Ghost fishing" by discarded or lost crab pots that crabs enter and cannot escape until the pot disintegrates or is retrieved is another problem. Challenges facing the soft crab industry nationwide include some of the same ones facing the hard crab industry, but an additional concern is the difficulty in obtaining peeler crabs for shedding purposes.[43]

Geographic boundaries set by humans are of no biological consequence to blue crabs, but political and economic considerations can affect the management, and therefore the survival, of the fishery. Such considerations occur for the Chesapeake Bay, which is surrounded by Maryland and Virginia. Each state has an interested legislature and a regulatory body to manage a resource that recognizes no political boundaries. In addition, since 1964, the bistate Potomac River Fisheries Commission manages the Potomac fisheries, except for the tributaries that are in each state.

Political realities call for careful coordination among the three management agencies. Although the fishery is not at the point that a crabber can run a trotline, even from an engine-powered boat, and be home with 1,800 hard crabs by ten in the morning, management initiatives in recent years seem to be maintaining a productive fishery in the Bay, albeit one, like most fisheries, with population fluctuations from year to year.[44]

Have Diminished Animal Abundances
Remodeled the Bay's Food Webs?

~~~~~~~~~~~~~

It is only necessary to bear in mind the enormous mass of these anadromous fish [in New England] one hundred years ago, and even later, to appreciate the influence they can exert in attracting fish from the outer waters to the shores and keeping them there for a considerable part of the year, and the lamentable result of the destruction of this source of supply, not only on its own account but also for its influence upon the sea fish. It is well known that while these anadromous fish were present there was an ample supply of cod, haddock, halibut, hake, and various other species close in to the shore . . . the fisherman, in an ordinary open boat, could go out and catch a full fare at a short distance from the land, both for use as fresh fish and for purposes of commerce, and that it was not until this source of supply was cut off that it became necessary to resort, to so great an extent, to distant parts of the sea.

Baird (1889)

LATE-NINETEENTH-CENTURY BIOLOGISTS LIKE Spencer Fullerton Baird understood that organisms were linked as prey and predators in what we now call food webs that involve flows of energy (calories) and nutrients among species. Baird (above) described how the human-caused decline of anadromous shad and river herring (prey) affected coastal populations of marine predators like cod, haddock, and others living in the Atlantic Ocean and thus changed the food web. This "economy of nature" was explained in more detail by Charles Stevenson (1899a):

The relationship between the different species of fish in the economy of nature is not very well understood, but sufficient is known to indicate that the valuable shore fisheries on the New England coast are intimately associated with the run of shad and similar species up the rivers of that section. Seventy years ago the run of fish up the rivers of the New England States was very much greater than at present, and after the parent fish had disappeared the waters swarmed with the young, which later in the year descended to the sea in enormous schools, attracting the cod, haddock, and other offshore species, which were caught in great abundance within a short distance of the coast, rendering unnecessary the expensive and hazardous trips to distant banks. But with the depletion of shad, alewives, salmon, and kindred species came a corresponding diminution in the number of cod, haddock, etc. near the coast. And it appears that any measures tending to restore the anadromous fishes to their former abundance will also improve the coast fisheries.

As described, the food web involved predators feeding on the enormous schools of anadromous prey swimming inshore to or from estuaries, so when the prey were captured by humans while coming or going, fewer adults and young were available to serve as prey for inshore predators. Fewer prey supported fewer predators, and the stocks of coastal predators declined inshore, requiring commercial fishers to sail farther out to sea.

The greatly simplified food web for the Chesapeake Bay in figure 9.1 focuses on the organisms mentioned in this book to introduce some sense of how those species interact. How has the depletion of the populations of formerly abundant species affected the Bay's food web and the web's functions? Are we past a point of no return so that we cannot return to an earlier, perhaps more productive, food web? Does that matter? The challenge facing anyone wishing to develop a colonial food web or a nineteenth-century food web to use as a model for Bay restoration is to estimate how changes in prey availability for the predators has affected the predators themselves. Let's start with a more detailed look at the role of anadromous fish and migrating waterfowl in the Chesapeake Bay food web.

What Happens When Predator or Prey Numbers Change?

Professor William Brooks described how shad participate in the Bay's food web in his 1893 report on fish and fisheries of Maryland:

> The supply [of shad] for the market is caught during the spring migration, when the fishes enter our inland waters heavy and fat after their winter's feast upon the abundant food which they find in the ocean. As they spend most of the year gathering up and converting into the substance of their own bodies the minute marine organisms which would otherwise be of no value to man, and as their instincts compel them to bring back to our very doors this great addition to our food supply, and thus put at our service a great and fertile area of the ocean, which, without their aid, would be beyond our control and of no value to man, their economic importance is very great.

In this paragraph, Brooks is relating how shad convert marine foodstuffs of no use to humans into shad tissue important as human food (this process also applies

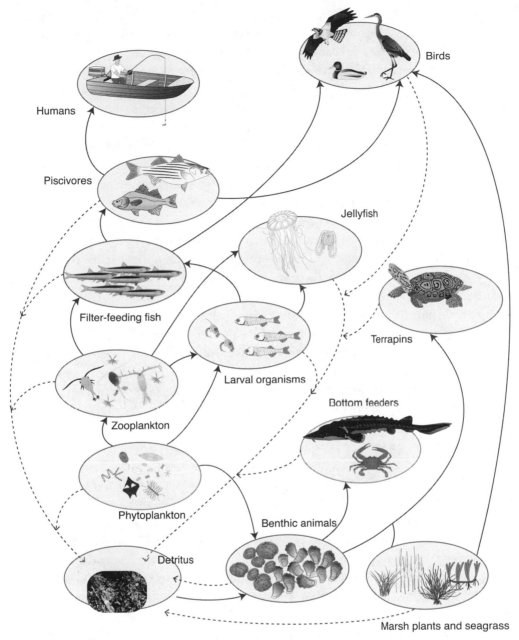

Figure 9.1. Simplified food web in the Chesapeake Bay with links between groups of organisms. Solid lines represent flows of food (energy) from one category to another; dashed lines show the path of decaying organisms to a material called detritus (mostly dead material). Drawing by Anne Gauzens.

to river herring and sturgeon). However, many estuarine creatures also take advantage of the influx of anadromous fish for their food. Should a juvenile or adult anadromous fish be eaten by a predator while in the estuary, or die before reaching the sea, the calories and nutrients in its body become available to the predator or a scavenger. The billions of eggs spawned by females are themselves packets of calories and nutrients that are eaten by many freshwater and estuarine creatures. Living fish produce feces, urine, and body mucus, material that becomes nutrients available to plankton in the water column. In a like manner, the waterfowl that descend on the Bay in winter transport calories and nutrients in their tissues from regions they had fed in recently. They too may become food for predators and scavengers, and they deposit their wastes in the Bay ecosystem while overwintering.

Scientists call this transfer of energy and nutrients from one region to another a *subsidy* that helps support the creatures that are resident in the system at that time.[1] Consider how such subsidies might have changed as a result of the decline in shad, river herring, and sturgeon from their multiple millions to levels now so diminished that the fisheries are closed. Estuarine piscivores (fish eaters) such as striped bass, white perch, blue fish, weakfish, eels, and channel catfish as well as fish-eating birds like herons, ospreys, eagles, cormorants, and some ducks now have fewer of these prey species, both juveniles and adults, to feed on from the spring through the autumn than in decades past. And it is not only the juveniles and adults that serve as food while in the estuary. The multimillions of eggs deposited and swarms of larvae hatched are also prey for organisms like zooplankton, shrimp, crabs, and small fish. Given the diminished number of spawning anadromous fish, there must be many, many fewer eggs and larvae available now in spring and early summer. What effect has the decline of eggs, larvae, juveniles, and adults had on their predators? Have the predators also been diminished because they have less food available, or have they switched to other prey not affected by a fishery?

Looking at the food web from another perspective, anadromous fish are themselves predators while in the estuary. While adult shad and river herring may or may not feed during their incoming and outgoing migration, their young feed on insects and zooplankton in the estuary before heading out to sea in the autumn.[2] Have the diminished numbers of these predacious young allowed their prey to thrive, perhaps to the benefit of other insect eaters or of predators like jellyfish that feed on zooplankton? Sturgeon use the sensory feelers or barbels on their lower jaw (see figure 7.1) to search the estuarine bottom for worms, snails, small clams, shrimp, and small fish. What has their decline meant for their former prey? In sum, have prey populations increased in the absence of anadromous predators? Have other species like small striped bass, croaker, or white perch taken advantage of the decline in anadromous species to eat the prey that those species once ate?

Some of the same questions can be asked about terrapins, which feed on bottom-dwelling animals and on marsh snails living on marsh grasses, and about waterfowl that feed on submerged grasses, clams, or worms. Distributions of geese and swans have shifted in recent decades from shallow aquatic areas of the Bay to agricultural fields where they feed on corn left after the autumn harvest. A change to an invertebrate diet by some ducks, including canvasbacks that now feed on clams, has been attributed to a decline in submerged vegetation that has occurred in the Bay.[3] How have such changes affected the flow of energy and materials within the food web?

The situation involving blue crabs is particularly intriguing. Although still abundant today, blue crabs seem to have been even more abundant in the past, as illustrated by the descriptions by Commissioner Hugh Smith in 1891 and Professor William Brooks in 1893 (see chapter 8). Blue crabs in the past seemed to some observers to be constantly hungry, presumably from intense competition for food. Thus this 1901 speculation from the Maryland Commissioners Pan-American Exposition: "During the season, from April to October, the shallower waters of the shores and estuaries of Chesapeake Bay, as well as the waters on the ocean side, contain an indescribable number of crabs. This abundance causes a fierce competition for food so that the crabs are always hungry and ready to seize any sort of animal bait."

This observation implies that blue crabs at that time were food limited, which in turn implies that the populations of clams, worms, and small fish were depleted by the foraging of ravenous crabs. Unfortunately, we have no data on the abundances of this crab food, so the effects of higher blue crab populations on their prey in the previous century are unknowable.

The ecological effects of diminished abundances of commercial species involve more than the loss of the bodies that could have been food for other members of the food web. I mentioned that living organisms deposit their body wastes in the ecosystem. What has been the effect of the diminished availability of the resulting nutrients from feces, urine, or mucus that serve as fertilizers for microbes and phytoplankton, which in turn are food for zooplankton, filter-feeding bivalves, and the like? The abundant waterfowl in the creek that kept Jasper Danckaerts awake in 1679 were doing more than "screeching"; they were defecating then, and also during the day. Their wastes provided nutrients that must have added a tremendous jolt of fertilizer to any body of water they visited in winter. What was the food web like when the "millionous multitudes" of birds left for the north and the waters warmed in spring, enabling bacteria to break down feces and release nutrients rapidly? Did great blooms of phytoplankton take advantage of the increased nutrients, or were the abundances of filter-feeding fish, oysters, clams, and other creatures able to keep the phytoplankton under control by grazing?

Did the Decline in Oysters Affect Jellyfish Numbers?

Some species may be so depleted by fisheries that they are essentially absent ecologically. That is, although the species might not be actually extinct, they are ecologically extinct, playing a minimal ecological role in the system they inhabit. John

Waldman calls such fish species "ghost fish."[4] Shad and river herring seem to fit in this category.

This concept of ecological extinction can apply to organisms other than fish. An example involves the oyster, whose populations have plummeted (see chapter 4). Oysters are filter feeders, pumping Bay water through their gills and extracting phytoplankton. The populations of oysters 150 years ago are estimated to have been able to pump a volume of water through their gills equal to the volume of the Chesapeake Bay (not all the water—just a volume equal to all the water) in three or four days.[5] Present numbers of oysters, being less than 1% of the numbers 150 years ago,[6] would be expected to take 100 times as much time, or 300 to 400 days, to pump the same volume of water. Thus, oysters now filter a tiny proportion of the available phytoplankton, leaving large amounts uneaten and available to support increased numbers of zooplankton (unless other filter feeders like clams, mussels, barnacles, some worms, and sea squirts have assumed the oysters' role; see figure 2.5).

The animal that is thought by some scientists to have benefited most from such a change in phytoplankton abundance, which in turn supports zooplankton abundance, is the sea nettle, a summertime pest that feeds on zooplankton. The decline in oyster abundances is thought to have shifted the Bay's food web from a benthic system dominated by animals living on the Bay bottom (predominantly oysters and other creatures using oyster reefs as a habitat) to a pelagic (water column) system dominated by zooplankton, including the larger gelatinous zooplankton like sea nettles that feed on smaller zooplankton (see figure 9.1 for the flows of food or energy).[7]

Finally, as mentioned in chapter 2, the Chesapeake Bay is an erosional or depositional system, with hard substrates being relatively uncommon and with reefs of oysters an important component of hard substrates. The shells of these ecological engineers provide an essential living space, not only for oyster spat but also for other creatures that attach to the shells (see figure 2.5). Also, mud crabs, other worms, and clams live in the sediment deposited among the shells. The great loss of oysters and their shells to the system must have meant that there was less living space for these hangers-on, many of whom are themselves filter feeders and who, with their eggs and larvae, serve as food for fish and crab predators that also use oyster reefs as habitat. We have limited understanding of the consequences of the diminished role of creatures that lived on and among oyster shell in the Bay's food webs.[8]

Understanding That Baseline Shift Can Inform Management and Restoration Activities

A recent article about long-term ecological change stated that managers need to understand how the biomass or abundance of species in an ecosystem has changed in order to manage the target ecosystem.[9] If managers and those restoring an ecosystem do not realize that some species have become ecologically extinct, that may hamper their management or restoration actions (depending on how ecologically important the diminished species was). Although the Chesapeake Bay in recent centuries has probably not lost any ecologically important species, the changes in abundances have to be taken into account in any attempt to describe how the Bay used to function in the 1800s. This is fertile ground for experts interested in modeling energy flow back when baselines were very much higher than they are now.

Afterword

For a number of reasons, including the loss or degradation of habitat caused by increased human population in the watershed, we will undoubtedly not be able to return the Chesapeake Bay to the productivity of the 1800s. Further, the anecdotes I recount in this book, given that they *are* anecdotes and not quantitative estimates of early animal populations, can serve mainly as a coarse guide rather than a clear path for restoration actions. As they are today, environments in the past were dynamic and changeable rather than static, so the anecdotes and the more quantitative reports represent just a snapshot in time.[1] Also, climate conditions have changed and, given global warming's onrush, will change even more from those of pre-Colonial times, affecting organisms and ecosystems. This book has not been written to provide a guide to restoring Colonial productivities and population abundances. Nevertheless, understanding the productivity that the system was once capable of supporting ecologically *and* economically may help us aim higher in our restoration efforts and encourage spending what is required in terms of human and monetary resources to make the ecosystem less polluted and more productive. We cannot restore what we do not understand.

How successful can attempts to restore ecosystems be? As an example from land systems, Jesse Ausubel proposed that a "Great Reversal" in harvests has occurred during the twentieth century in many world forests.[2] Tree harvests in the United States peaked in 1906, declining thereafter, and the amount of forest began increasing, especially after 1950. Further, "studies of forest biomass for the 1990s in the boreal and temperate region in more than 50 countries show the forests expanding in area and/or volume." Ausubel suggested that the Great Reversal was thus becoming a "Great Restoration" whose principles could be applied to terrestrial nature in general.

What then of restoring marine resources in general and in the Chesapeake Bay in particular? First, let us focus on marine resources in general. In 1971, New Zealand passed a Marine Reserves Act, and in 1975 their first reserve was created at Cape Rodney–Okakari Point in the North Island.[3] The reserve was so successful that there are now 44 reserves in New Zealand's territorial waters. Fishing is banned in reserves, but recreational activities (kayaking, sailing, snorkeling, diving) and scientific studies are permitted and thrive. Local fishers were initially unhappy with the Cape Rodney Reserve being established, but that attitude has changed. Numerous commercial species of fish and shellfish flourished within the reserve after fishing ceased, and the overharvested ecosystem refreshed itself. The increasing abundances of animals within the reserve spilled over into the adjacent habitat where commercial species can be fished.[4] As a result, harvests outside the reserve are commercially remunerative, and fishers are happy. Similar spillovers of protected organisms come from other reserves worldwide, demonstrating the resilience of marine systems safe from fishing or other disturbance.

Second, let's turn to the Chesapeake Bay in particular to examine striped bass (stripers) and oyster restoration. In the 1970s and early 1980s, striper populations declined from about 15 million pounds in 1973 to about 2 million in 1983, causing concern among managers and fishers. In 1985, Maryland banned fishing for stripers, and Virginia followed in 1989. A hatchery program provided young fish to supplement natural productivity.[5] Studies revealed important insight into striper biology that improved management techniques. The striper population was restored to sustainable levels, and the fishery was reopened.

Great effort is being put into building up oyster reefs in selected regions around the Bay, with federal, state, and nongovernmental agencies working together.[6] Reefs are having their profiles restored, and increased spat settlement is expected. Time will tell, but oysters are a resilient species, so hopes are high. As New Zealand's reserves benefit adjacent fisheries, oyster sanctuaries in the Bay can help repopulate nearby depleted oyster reefs by producing larvae that are carried by currents outside the sanctuaries to those reefs. Indeed, recent research in North Carolina has demonstrated that oyster reserves in Pamlico Sound may produce 4 to 700 times more larvae per square meter of reef than do nearby harvested reefs, thus potentially subsidizing the harvested reefs.[7]

History tells us that humans can have a deleterious effect on natural resources if harvests are not carefully managed and if environments, including spawning habitat, are allowed to deteriorate. History also tells us that an understanding of how organisms fit into their ecosystem and how ecosystems work can allow restoration projects to be successful. It is reasonable to hope that the Chesapeake Bay's commercial and recreational fisheries can be restored to a higher level of sustainable harvest, even if never to attain the harvest levels experienced by nineteenth-century watermen. With informed action we can shift baselines to higher levels if we have the resolve and apply the appropriate human and monetary resources.

Appendix

Fishing Gear and Methods

Betwixt their hands and thighes, their women use to spin, the barks of trees, deare sinews, or a kind of grasse they call Pemmenaw, of these they make a thread very even and readily. This thread serveth for many uses . . . as also they make nets for fishing, for the quantity as formally braded as ours. They make also with it lines for angles. Their hookes are either a bone grated as they nock their arrows in the forme of a crooked pinne or fish-hook, or of the splinter of a bone tied to the clift of a little stick, and with the end of the line, they tie on the bate. They use also long arrowes tyed in a line, wherewith they shoote at fish in the rivers. But they of Accewmack use staves like unto Javelins headed with bone. With these they dart fish swimming in the water. They have also many artificiall weares in which they get abundance of fish.

Smith (1612)

In their quest to capture food from the Chesapeake Bay, humans have used various devices, ranging from the simple to the sophisticated.[1] Under conditions of limited technology, many devices were made from local resources, as nicely demonstrated by the 1590 account of Indian fishing gear and methods by Thomas Hariot in the frontispiece to this book and by Smith's description above (see also Meehan's 1893 statement on page 23). Most of the Indian devices described by Smith, including hook and line fishing and spear fishing, were familiar to the settlers and will be to Bay residents today. Other devices are less familiar, so I describe them below.

Shad, River Herring, and Sturgeon

Three types of nets were used in the Chesapeake Bay to capture shad, river herring, and sturgeon, as well as other commercial fish species. The simplest was the gill net, a swath of mesh netting hanging vertically in the water to capture fish that blunder into it (figure A.1). The fish's forward momentum forces its head through a mesh, and its flared gills become trapped so it cannot back out. Gill nets can be deployed to float freely in the water, requiring constant attention from fishers as they drift with the current. In the figure, a man in the late 1800s pays out a net attached to a lantern as two colleagues row across the river channel. When the net is completely paid out, it is attached to a second lantern, and the men then move continuously back and forth along the net to remove gilled fish. Alternatively, gill nets were attached to stakes embedded in the river bottom and left to fish for a time without supervision. Nets of either design were relatively inexpensive to knit and could be handled by one or more fishers with a rowboat.

A much more expensive net was the haul, or beach, seine.[2] These were deployed in a semicircle to surround fish, with the net hauled onto the beach along with the trapped fish

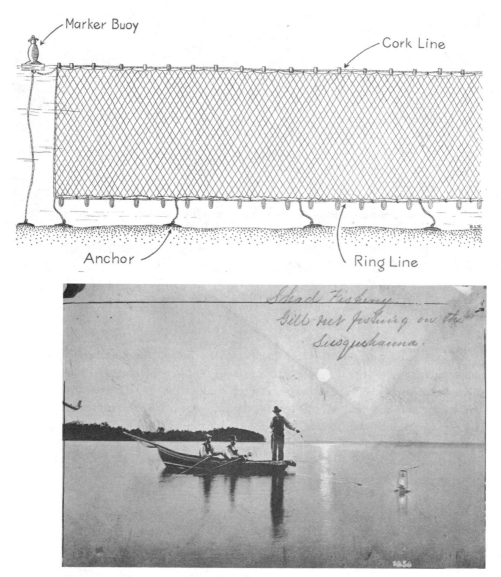

Figure A.1. Gill net *(top)*; gill-net fishing *(bottom)* on the Susquehanna at night. Coker 1949; Goode 1887, pl. 163 NOAA Photo Library.

Figure A.2. Huge beach seine being hauled on shore at Stony Point, Virginia, in the late 1800s. NOAA Photo Library.

(figure A.2). Seines could be 6 to 15 feet deep and miles long, depending on the depth and size of the river being fished. Large-haul seines needed a crew of many men to deploy and retrieve them; very large seines required horses or steam engines to haul them onto the shore. This infrastructure added to the cost of a seining operation. Here is a description of a fishery by journalist Anne Royalle in 1826:

> In the first place, about fifteen or twenty men, and very often an hundred, repair to the place where the fish are to be taken, with a seine and a skiff. This skiff however, must be large enough to contain the net and three men—two to row and one to let out the net. These nets, or seines, are of different sizes, say from two to three hundred fathom [1,200 to 1,800 feet] in length, and from three to four fathom wide [deep]. On one edge [top] are fastened pieces of cork-wood, as large as a man's fist, about two feet asunder; and on the opposite edge [bottom] are fastened pieces of lead, about the same distance—the lead is intended to keep the lower end of the seine close to the bottom of the river. The width of the seine is adapted to the depth of the river, so that the corks just appear on its surface, otherwise the lead would draw the top of the seine under water and the fish would escape over the top.
>
> All this being understood, and the seine and rowers in the boat, they give one end of the seine to a party of men on the shore who are to hold it fast. Those in the

boat then row off from the shore, letting out the seine as they go; they advance in a straight line towards the opposite shore until they gain the middle of the river, when they proceed down the stream until the net is all out of the boat except just sufficient to reach the shore from whence they set out, to which they immediately proceed. Here an equal number of men take hold of the net with those at the other end, and both parties commence drawing it towards the shore. As they draw, they advance towards each other, until they finally meet . . . when the fish begin to draw near the shore, one or two men step into the water, on each side of the net, and hold it close to the bottom of the channel, otherwise the fish would escape underneath. All this being accomplished, the fishermen proceed to take out the fish.

Not every fishing group had access to a wide and snag-free beach. Consequently, in the late nineteenth century a number of anchored shad floats were used in the Susquehanna as an alternative landing place. A typical float was built of floorboards over logs, perhaps 50 to 100 feet wide and 200 to 300 feet long.[3] Large floats could hold a bunkroom and mess hall for 20 to 100 men, as well as sheds for gutting and salting fish and for holding them in barrels. Given the effort required to pull the huge seine and its catch towards the float, in early times a horse was used to turn a winch, so the float would have a stable until steam engines replaced horses. A wooden platform attached to the downstream edge of the float at an angle to the river bottom served as a platform up which the net with its catch was dragged (figure A.3, *top*). The whole assembly was moored upstream of the river bottom that was to be swept by the seine, one end of which was attached to the float. The captured fish (figure A.3, bottom) were sorted by species and size, then moved to the processing sheds or discarded. The amount of capital that must have been invested in building and deploying such a complex structure gives some idea of how remunerative the fishery was. I have never found out where shad floats were moored during the rest of the year after the short spring fishery ended.

A fixed net of complex construction was the pound net, or weir (figure A.4). These are still used to capture fish in the Bay, but their numbers are very much fewer than when they were being used to capture shad and river herring (see chapter 3). Brooks described a pound in 1893:

The pound [head] is a large enclosure of a very complicated pattern, shut in by a fence, which is formed of strong netting stretched upon piles, or posts, which are firmly planted on the bottom. The lower edge of the wall of netting rests on the bottom, while its upper edge is high enough above the surface of the water to keep the fish from escaping by jumping over it at high water.

A straight wall of netting runs out from the shore [the leader] and turns the fishes which run against it out into the deep water, where it ends just inside the opening into the first or big heart. This opening is about twenty-five feet wide, and it is so large that the fishes enter it fearlessly and swim about until they are stopped by the wall of the heart, when, in their efforts to escape into deep water, they are gradually guided by the walls into the inner [second] heart, and from this, through a narrow opening [funnel] only a yard wide, into the pound.

This is a rectangular trap, about fifty feet wide, with its bottom, as well as its sides, covered with netting. The bottom is weighted around its edge by sinkers of lead, and it is kept stretched and flat by means of lines, which pass through metal

Figure A.3. (Top) Preparing a haul seine on the deck of a Susquehanna River shad float. Note the wooden slope to the left up which the seine will be dragged, the cork floats of the top of the seine and the bottom lead line, and the buildings to the right. *(Bottom)* Fish from a seine hauled onto the wooden deck of a shad float near Port Deposit, Maryland, in 1905. Courtesy Calvert Marine Museum, Solomons, Maryland; Courtesy Historical Society of Cecil County, Elkton, Maryland.

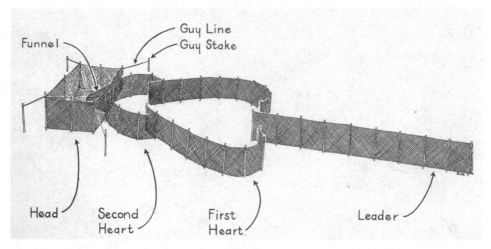

Figure A.4. Fish pound net diagram. Coker 1949.

rings at the bottoms of the posts, and are then made fast above water. The netting is so arranged that it may be detached from the posts by the fishermen in their boats, and gradually raised to the surface until all the fishes are drawn together and penned in one corner, where they may be dipped out of the water with hand nets.

A pound being fished by pulling the net to the surface where the trapped fish could be bailed out into a boat is shown in figure A.5.

Oysters

Gear used in the early oyster fishery in the Chesapeake Bay (figure A.6) was simple and easily made by blacksmiths. An anonymous author wrote in 1869:

> The implements used in oyster fishing are few and simple in construction. They are the dredge [and] the tong . . . The dredge is used on the natural beds, in deep water. It is an iron net set in pear-shaped iron frames, and furnished with teeth so arranged as to tear the oysters from the beds, and gather them into the net as it is drawn over the bottom by the vessel, to which it is attached by means of a long rope. It weighs about one hundred and fifty pounds, and is drawn on board the vessel by a windlass arranged for the purpose. It is designed to hold about three bushels, though it is rarely filled with marketable oysters at one "haul." When one-fourth of the contents is good oysters, the "haul" is considered a good one. The remaining empty shells are cast back into the water. The tongs are composed of two iron rakes attached to long wooden poles, with an axle set near the rakes. The fisher leans over the side of his boat, and handles this tool with ease in water from two to eight feet deep.

The basic designs shown in figure A.6 have changed little since 1869. Dredge boats came to deploy two dredges, one port and one starboard. Each was raised to the surface by a winder with two to four handles (two in the figure) turned by men. Steam winches came later. Hand tongs were deployed from smaller boats. The length of tonging shafts depended on the depth of the tonging bed being harvested, becoming longer as shallow oysters were depleted.

Smaller tongs or nippers had fewer teeth and were used in shallow water to pick up single oysters or small clumps seen from the surface. In 1887, Charles Marsh of Solomons, Maryland, invented "patent" tongs used in water beyond shaft tong depth that were closed by a lever system of ropes attached to a mechanical winch on board the tonging boat. The ropes pulled up on two metal arms, closing the tongs on the bottom (in the twentieth century, a hydraulic system was developed to close and retrieve the patent tongs).

Terrapins

In addition to baited hooks, dip nets, and seines (see chapter 5), traps called fyke nets (figure A.7) used to capture terrapins were anchored in shallow water (they are also used to capture fish today). They incorporate a number of nested hoops supporting nets and funnels of strong cotton. In the Chesapeake Bay, hoops were usually three feet in diameter and the device about four feet long, but dimensions were different in other regions.[4] The trap could be baited with fish and tied to a pole in such a way that a portion projected into the air to allow captured terrapins to breathe.

Figure A.5. Pulling up the net inside a pound in the Potomac River, to raise captured fish to the surface. NOAA Image Library.

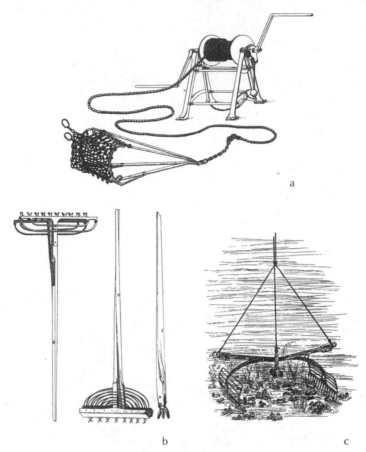

Figure A.6. *a*) Oyster dredge retrieved by a winder with two handles; *b*) Three hand tongs with a toothed iron basket *(left),* wooden basket *(middle),* and small "nippers" *(right)* attached to the ends of wooden shafts; *c*) "Patent tongs" closed by a mechanical lever system. Goode 1887, pl. 237, 238; Smith 1891b, pl. XLIV.

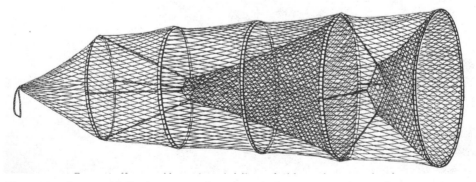

Figure A.7. Fyke net with five hoops, two funnels, and a loop at one end to anchor to a stake. Coker 1949.

Crabs

Crabbers seeking soft crabs and peeler crabs towed two or three simple scrapes through seagrass beds from a sailboat (figure A.8). The 2.5-foot-wide scrape resembled a light oyster dredge but had no teeth on the front bar, and the 4-foot-deep bag was cotton, not metal.[5]

Hard crabs were fished with trotlines of various lengths, baited at regular intervals and with buoys at each ending marking the trotline's location (figure A.9). When sail was used, the line was deployed directly across the wind direction so that the crab boat could run between the buoys without tacking.[6] As the boat proceeded, the trotline ran up from the estuary bottom, over an outrigger roller, and back into the water. A wire-mesh basket was used to capture hungry crabs holding tightly to the bait as it neared the surface.

Figure A.8. Crab scrape in sea grass. Cargo 1954.

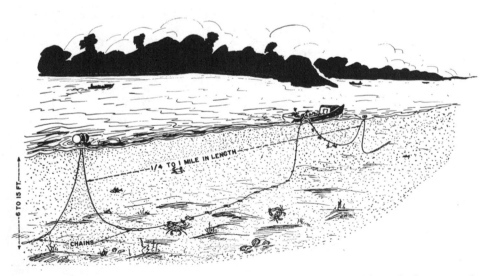

Figure A.9. Crab trotline with a float and weights at each end, being raised from the bottom and run over a roller on the boat's gunwale to bring crabs attacking the bait to the surface where they are dip-netted. Cargo 1954.

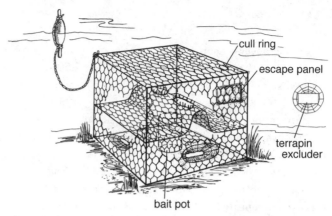

Figure A.10. Wire mesh crab pot. Courtesy Maryland Sea Grant, from their publication by Kennedy, V.S., and L.E. Cronin, 2007. *The Blue Crab:* Callinectes sapidus. Redrawn by Deborah Kennedy from a figure in Stehlik, L.L., P.G. Scarlett, and J. Dobarro, 1998. Status of the blue crab fisheries in New Jersey. *Journal of Shellfish Research* 17:475–485.

In the later twentieth century, crabs were also captured in wire-mesh pots with four entrance funnels, a hinged lid for access to the catch in the upper compartment, and a bait holder in the middle (figure A.10). Pots today must have an excluder device in the entrance funnels to keep terrapins out while allowing crabs entry. An oval cull ring lets out undersized crabs. An escape panel is held in place with metal attachments that will eventually corrode if the pot is abandoned (a ghost pot), allowing the panel to fall away, leaving an opening that lets crabs that enter the ghost pot escape.

Further Reading

Brooks, W.K. 1996. The Oyster. Johns Hopkins University Press, Baltimore, MD. 230 pages.
A reissue of Brooks's 1891 report on the oyster and its declining Maryland fishery.

Brush, G. 2017. Decoding the Deep Sediments: The Ecological History of Chesapeake Bay. Maryland Sea Grant Publication UM-SG-CP-2017-01, College Park, MD. 63 pages.
Compares land use in pre- and post-Colonial periods as reflected by sediment profiles in bottom cores, and relates land-use changes to the aquatic health of the Bay.

Curtin, P.D., G.S. Brush, and G.W. Fisher. 2001. Discovering the Chesapeake: A History of an Ecosystem. Johns Hopkins University Press, Baltimore, MD. 385 pages.
Comprehensive overview of the Chesapeake ecosystem covering climate, terrestrial, and aquatic subsystems and the influence of Indians, early settlers, and modern inhabitants.

Dodds, R.J., and R.J. Hurry. 2006. It Ain't Like It Was Then: The Seafood Packing Industry of Southern Maryland. Calvert Marine Museum, Solomons, MD. 95 pages.
Extensive illustrated history of large and small seafood packing firms in southern Maryland counties.

Ernst, H.R. 2003. Chesapeake Bay Blues: Science, Politics, and the Struggle to Save the Bay. Rowman & Littlefield, Lanham, MD. 205 pages.
A political scientist comments on attempts by environmentalists, managers, scientists, the seafood industry, and politicians to conserve the Chesapeake Bay.

Franklin, H.B. 2008. The Most Important Fish in the Sea: Menhaden and America. Island Press / Shearwater Books, Washington, DC. 280 pages.
An extensive overview of an ecologically and commercially important fish in the Chesapeake Bay.

Gerstell, R. 1998. American Shad in the Susquehanna River Basin: A Three-Hundred-Year History. Pennsylvania State University Press, University Park, PA. 217 pages.
A detailed illustrated history of the Susquehanna shad fishery based on tax and land records.

Hay, J. 1979. The Run, 3rd Edition. W.W. Norton, New York. 184 pages.
A popular account of alewives and their annual migration up a Cape Cod river.

Hersey, J. 1988. Blues. Vintage Books, New York. 208 pages.
A meditation on fishing, especially for bluefish, a predatory species that occurs regularly in the Chesapeake Bay.

Houde, E.D. 2011. Managing the Chesapeake's Fisheries: A Work in Progress. Maryland Sea Grant, College Park, MD. 126 pages.

Considers ecosystem-based management in place of species-specific management of commercial fisheries.

Jackson, J.B.C., K.E. Alexander, and E. Sala (Editors). 2011. Shifting Baselines: The Past and the Future of Ocean Fisheries. Island Press, Washington, DC. 296 pages.
Describes changing abundances of selected fisheries worldwide.

Jacoby, M.E. 1991. Working the Chesapeake: Watermen on the Bay. Maryland Sea Grant, College Park, MD. 156 pages.
Stories of watermen and their boats and gear.

Johnson, P. (Editor). 1988. Working the Water: The Commercial Fisheries of Maryland's Patuxent River. University Press of Virginia, Charlottesville, VA. 218 pages.
Illustrated history of commercial fisheries in the Patuxent River, a microcosm of Bay fisheries over the years.

Lang, V. 1961. Follow the Water. John F. Blair, Winston-Salem, NC. 222 pages.
First-hand description of oystering, crabbing, clamming, fishing, and hunting by a man who left university teaching to become a professional waterman in Talbot County, Maryland.

Lippson, A.J., and R.L. Lippson. 2006. Life in the Chesapeake Bay, 3rd Edition. Johns Hopkins University Press, Baltimore, MD. 324 pages.
Nicely illustrated compilation of information accessible to lay readers.

McPhee, J. 2002. The Founding Fish. Farrar, Straus, and Giroux, New York. 358 pages.
The natural history of American shad, their migrations, their fishery, and their role in US history. Rebuts the "providentialist canard" that the appearance of shad in the Schuylkill River saved Washington's starving troops at Valley Forge in 1778.

Meanley, B. 1975. Birds and Marshes of the Chesapeake Bay Country. Tidewater, Centreville, MD. 157 pages.
Natural history accounts of Bay birds, including waterfowl, and their estuarine habitat.

Russell, D. 2005. Striper Wars: An American Fish Story. Island Press / Shearwater Books, New York. 368 pages.
The story of attempts to restore striped bass populations along the western Atlantic coast after a steep decline in the early 1980s.

Sullivan, C.J. 2003. Waterfowling on the Chesapeake, 1819–1936. 2003. Johns Hopkins University Press, Baltimore, MD. 195 pages.
Illustrated story of gunning clubs, shooting gear, Chesapeake Bay retrievers, decoys, and decoy carvers.

Waldman, J. 2013. Running Silver: Restoring Atlantic Rivers and Their Great Fish Migrations. Lyons Press, Guilford, CT. 304 pages.
Historical overview of the migrations of shad, herring, sturgeon, and salmon up rivers to spawn and the reasons for their declines.

Walsh, H.M. 1971. The Outlaw Gunner. Tidewater, Centerville, MD. 178 pages.
Illustrated history of market gunning in the Chesapeake Bay.

Warner, W.W. 1976. Beautiful Swimmers: Watermen, Crabs and the Chesapeake Bay. Little, Brown, and Company, Boston, MA. 304 pages.
Pulitzer Prize–winning account of the blue crab and its fishery and fishers.

Wennersten, J.R. 2007. The Oyster Wars of Chesapeake Bay, 2nd Edition. Eastern Branch Press, Washington, DC. 164 pages.
Story of the early feuding between Maryland and Virginia oystermen and between oystermen and managers.

Wharton, J. 1957. The Bounty of the Chesapeake: Fishing in Colonial Virginia. University Press of Virginia, Charlottesville, VA. 78 pages.
Uses extracts of Colonial writings to describe early Bay fisheries.

Notes

Preface

1. Census of Marine Life: www.coml.org; HMAP program: www.coml.org/projects/history-marine-animal-populations-hmap. Nontechnical summaries of HMAP studies: Starkey et al. (2008).
2. Virginia's oyster fishery: Haven et al. (1978).
3. Maryland's oyster fishery: Kennedy and Breisch (1983).
4. Chesapeake Bay oyster fishery: Kennedy (1989).
5. History of Chesapeake Bay commercial species: Kennedy and Mountford (2001); blue crabs: Kennedy et al. (2007); diamond-backed terrapin: Kennedy (2018).
6. South Australia oysters: Alleway and Connell (2015); Alleway et al. (2016).

A Note on Anecdotal and Quantitative Harvest Statistics

1. Fishery statistics in 1840: Van Metre (1915).
2. US Fish Commission: https://celebrating200years.noaa.gov/rarebooks/fisheries/welcome.html.
3. The Fisheries and Fishery Industries of the United States: Goode (1887).

Chapter 1. Shifting Baselines in the Chesapeake Bay, the Immense Protein Factory

1. Tidewater Indians ate well: Potter (1993).
2. Indian fishing tools: Rau (1884).
3. Indian subsistence patterns: Schindler (2008).
4. Small population sizes: Gallivan (2012).
5. Rich food resources: Scharf (1879).
6. George Alsop (1666) also decried the constant diet of venison (note that quotations in this book from old accounts keep their original grammar and spelling).
7. Many newspaper reports, magazine articles, and books in the late 1800s claimed that early laws in Maryland limited the number of days a week that terrapins or wild ducks could be served to servants or slaves, who protested this repetitive diet. An anonymous author writing in 1896 and Charles Stevenson in 1909 thought these stories were probably false, and I have found no such laws. Nonetheless, Bruce Leffingwell wrote in 1890 that a Chicago lawyer told him that when the lawyer owned a Chesapeake farm he hired slaves from their owners and fed them on canvasback ducks daily in winter until the

slaves rebelled and demanded pork. If not laws against monotonous diets, there may have been protests.

8. Shifting baselines: Pauly (1995).
9. Bison: Dodge (1877); passenger pigeons: Wright (1910).
10. Surprisingly, the first colonists apparently arrived in the Chesapeake region with limited or no fishing gear. Thus they differed from colonists in Canada and New England who established vigorous fisheries (Wharton 1957, Waldman 2013). Johnson (1908) attributed the difference between the New England and middle Atlantic fisheries in colonial times to the "limited possibilities of agriculture in New England and the wealth of life in the ocean that washes her shores." Agriculture around Chesapeake Bay was an easier prospect than in New England in Colonial times, but the Bay fisheries would eventually come into their own.
11. A common misconception is that the word "Chesapeake" means "Great Shellfish Bay" in Virginia Algonquian. Blair Rudes, an expert in the dialect, stated that the word probably means "Great Water" (Farenthold 2006).
12. In commenting on Burnaby's (1775) statement about 5,000 fish caught in a single seine haul, Spencer Fullerton Baird stated in 1889 that "it is probable that the seines used in the Potomac waters over a hundred years ago were much smaller than those now employed, one of one hundred yards being, doubtless, of remarkable magnitude." Thus a harvest as large as 5,000 in a short haul foretold the greater harvests that could be made in the 1800s with much longer nets (see chapter 3).
13. Road building: Stover (1987); www.fhwa.dot.gov/infrastructure/bankroad.cfm.
14. Roads for short-distance deliveries: Gallatin (1808).
15. Canals better than roads: Livingood (1941), Dilts (1993).
16. Baltimore lacks a "decent river": Dilts (1993).
17. C & O Canal: Dilts (1993).
18. Canals costly: Rubin (1961), Dilts (1993).
19. Canal or railroad?: Rubin (1961).
20. Railroad idea: Ringwalt (1888), Stover (1987).
21. B & O Railroad completed: Rubin (1961), Stover (1987), Dilts (1993).
22. Oyster shipments increase: Nichol (1937).
23. Kensett, Sr., first used cans in United States: Mayer (1871).
24. Wright cans oysters in Baltimore, followed by Kensett, Jr., and others: Judge (1895), Nichol (1937).
25. Baltimore canned goods in demand worldwide: Mayer (1871), Stevenson (1899b), Collins (1924), Kee (2006).
26. Efficiency of canning oysters: Stevenson (1894).
27. Seasonality of canning: King (1875).
28. New England branch plants: Sweet (1941); increase in processors: Van Metre (1915), Nichol (1937). Chapter 14 in Schultz (1908) provides a history of the tin can and its use in Baltimore.
29. First statistics in 1840: Van Metre (1915).

Chapter 2. Why the Chesapeake Bay Was So Productive and What's Changed

1. Detailed timeline: Chesapeake Bay Fisheries Ecosystem Advisory Panel (2006).
2. Geological history: https://pubs.usgs.gov/fs/fs102-98/.
3. Statistics on the Bay: Malone et al. (1999). See also www.chesapeakebay.net/discover/facts.
4. Number of tributaries: www.chesapeakebay.net/faq/keywords/rivers_and_streams#inline.
5. Flushing rate: Malone et al. (1999).

6. Estuarine retention: Odum (1970).
7. Sediments eroded: Hobbs et al. (1992).
8. Colonists and sedimentation rates: Cooper and Brush (1993), Cooper (1995), Brush (2017).
9. Sediment from farmland: https://earthobservatory.nasa.gov/IOTD/view.php?id=88523.
10. Nutrients: Kemp et al. (2005).
11. Increase in nutrients: Boynton et al. (1995).
12. Increased anoxia and hypoxia: Officer et al. (1984).
13. Food-web relations: Flemer et al. (1983), Kemp and Boynton (1992).
14. Overview of human-induced changes: Curtin et al. (2001).
15. Population growth: Chesapeake Bay Program (2012).
16. Wastes and water quality: Kemp et al. (1999), Boesch (2006).
17. Clean-up agreements and directives: Boesch et al. (2001).
18. Overview of "surges of development": Perez (2007).
19. Canals supplement roads: Livingood (1941).

Chapter 3. The Spring Fishery for Shad and River Herring

1. Shad, *Alosa sapidissima*, are herring-like anadromous fish. River herring include two other species of herring-like fish that arrive in Chesapeake Bay around the same time as the shad. Alewives, *Alosa pseudoharengus*, are also known as branch herring (they swim up into the smallest of tributaries or river branches) or spring herring (they arrive a short time before the shad). Blueback herring, *Alosa aestivalis*, are also known as summer herring (they arrive later than shad or spring herring) or glut herring (in the past they arrived in such numbers as to flood an already saturated fishery that had been harvesting shad and alewives for a few weeks).
2. Spawning: Brooks (1893a).
3. Indians eating shad: Gay (1892).
4. Indian names: McPhee (2002).
5. Washington's description of his shoreline: Sabine (1853); more information on his fishing practices: Wharton (1957), Tilp (1978).
6. Anthropological evidence: Wright (1884).
7. Susquehanna fisheries: Holberton (1892).
8. Binghamton, NY: Stevenson (1898).
9. Small seine sizes: Wilkinson (1840); later large sizes: Fitzgerald (1895), Gerstell (1998).
10. Shad floats: Fitzgerald (1895); Gerstell (1998).
11. Shad fishery importance: Smith (1895b).
12. Industrial dams: Walter and Merritts (2008).
13. Maze of fishing gear: Vincent and Downey (1903), Van Metre (1915).
14. See Limburg and Waldman (2009) on the decline of migrating fish.
15. Briefly, carrying capacity is the ability of an environment to support a species in that environment over time. It depends on factors like appropriate food and habitat and the presence or absence of other species that might compete for resources.

Chapter 4. The World's Greatest Oyster Fishery

1. Ecosystem engineers: Jones et al. (1994)
2. Oyster song: Frank Leslie's Popular Monthly for 1886, Volume 31(1):95.
3. De Bry's 1590 illustration of Native Americans fishing shows an Indian standing in a canoe and wielding what seems to be a rake. Was he oystering in shallow water? Note the reports at the head of chapter 4 on oyster reefs extending to the surface of the Bay and being navigational hazards. John R. Page reported to the Virginia Fish Commissioner in

1877 that "oyster rocks" grow "generation piled on generation, until they almost form islands in the rivers." Taking oysters from such shallow reefs would have been simple.

4. Oysters of the James: New York Times, May 16, 1880.
5. Dredging accident: Wyman (1884). Wyman gives interesting details about the design and workings of dredging schooners, and the hardships associated with dredging.
6. Accidents summarized: Stevenson (1894).
7. Oysters thrive: Alford (1975).
8. Large shells: Anonymous (1869).
9. This chapter focuses on Maryland's fishery. Virginia's fishery also declined, but Virginia watermen were not as adamantly opposed to leasing as in Maryland, and attitudes about private versus public reefs were different. Extensive overviews of Virginia's fishery are presented by Haven et al. (1978) and Schulte (2017). For a well-illustrated history of the North American oyster industry, see MacKenzie (1996).
10. In the early 1900s, a Maryland bushel held 329 marketable oysters on average: Mitchell et al. (1907). Fewer oysters may have filled a bushel in the 1800s: Stevenson (1894) wrote "Compared with the conditions of thirty-five years ago . . . because of the very vigorous fishery to which they have been subjected the size of the oysters brought to market is less." That situation seems to have persisted: when Varley Lang wrote in 1961, it took about 400 oysters to fill a bushel, presumably because there were fewer large oysters available than in the early 1900s.
11. Oyster dredging data: Mitchell et al. (1907).
12. Commander William Timmons of the Fishery Force of Maryland gave reasons (1874) for thinking that dredging was benign and that oyster populations were not declining.
13. Statistics about Baltimore's oyster industry in 1869 by Anonymous (1869) were based on remarks by Thomas Kensett to the Baltimore Oyster Packers Association in April 1869 (Mayer 1871); see also Howard (1873) and Schultz (1908) on Baltimore's canning industry.
14. Skipjack numbers: https://oystercatcher.com/skipjack-history/ and http://lastskipjacks .com/index.html.
15. 2014 data in National Marine Fisheries Service (2016).
16. More details on historical legislation that supplement and extend Grave's graph appear in Kennedy and Breisch (1983).
17. The 1869 Commissioner of Agriculture's Report includes a statement on oysters by an anonymous writer.
18. Early dominance of the US oyster industry: Kennedy and Breisch (1983); Keiner (2009).
19. Overfishing in New England: Ingersoll (1881), Van Metre (1915), Sweet (1941).
20. Fair Haven: Nichol (1937).
21. Relocating a fishery from a depleted region (e.g., New England) to a less depleted region (e.g., New Jersey and Virginia) is common in fisheries worldwide and is referred to as "serial depletion."
22. Dredge bans in Virginia (1811) and Maryland (1820): Stevenson (1894).
23. More regulations: Stevenson (1894), Grave (1912), Schulte (2017).
24. Maryland State Fishery Force: Earle (1932), Wennersten (2007).
25. Virginia marine police: Schulte (2017).
26. Fourteen million bushels: Grave (1912).
27. Tangier and Pocomoke Sounds surveyed: Winslow (1881, 1882, 1884).
28. Oyster Commission Report: Brooks et al. (1884).
29. Popular treatise: Brooks (1891, 1905); reissued as Brooks (1996).
30. Fifteen million bushels: Stevenson (1894).
31. Oyster Commission recommendations undercut: Grave (1912).

32. Leasing Virginia oyster reefs: Schulte (2017).
33. Maryland's cull law: Stevenson (1894).
34. Cull law is unpopular: Brooks (1891).
35. Decline in Maryland's industry: Van Metre (1915), Nichol (1937).
36. Leasing fees to be used for public infrastructure: Haman (1893).
37. Editorial support for leasing: e.g., Baltimore Sun (1893, 1903).
38. Haman Bill: Kennedy and Breisch (1983).
39. Virginia oyster bed survey by Baylor (1895); Maryland survey by Yates (1913).
40. Extensive reviews: Ingersoll (1881) and in Maryland: Stevenson (1894).
41. Sociopolitical resistance by oystermen: Maltbie (1914), Green et al. (1916).
42. Canning industry employment over time: Nichol (1937).
43. Legal challenges: Christy (1964), Power (1970).
44. History of Shepherd Bill: Kennedy and Breisch (1983).
45. Public versus private oyster bed yields: Bowman (1948), Hammer (1948).
46. Committee recommendations ignored: Kennedy and Breisch (1983).
47. Political considerations affect fisheries: Wallace (1952).
48. Disease in the Bay: Andrews (1979).
49. Aquaculture accelerating: Kobell (2017).

Chapter 5. Diamond-backed Terrapins

This chapter, which focuses on the Chesapeake Bay fishery, includes information extracted from a description of the wider US fishery for terrapins in Kennedy (2018).

1. Turtles in middens: Sutherland (1974), Thorbjarnarson et al. (2000), Handley (2001).
2. Terrapins fed to pigs: Laffan (1877).
3. Clayton's oxcart: New York Sun (1904).
4. Old Bay Line dollar menu: Brown (1940).
5. Taste preferences for Chesapeake terrapins: Coker (1906); for Potomac River or Chester River terrapins: Baltimore Gazette (1887).
6. Maryland's harvest ban: Bay Journal (2007).
7. Progging: True (1887), Coker (1906).
8. Dredging: Ducatel (1837), True (1887), Brooks (1893a).
9. Rapping attracts terrapins: Thom (1898).
10. Seining: True (1887), Brooks (1893a).
11. Trapping females: New York Sun (1904).
12. Inflation calculator: http://westegg.com/inflation/.
13. Market expansion: Baltimore Gazette (1877).
14. Baltimore the greatest terrapin market: New York Times (1886).
15. Food for penned terrapins: Baltimore Gazette (1877), New York Times (1888).
16. Damage to penned animals: New York Times (1888), Lowry (1888).
17. Prices escalate: Laffan (1877), Washington Star (1884), New York Times (1888), Baltimore Sun (1894), New York Sun (1904).
18. Terrapin dishes adulterated: Baltimore Sun (1887), New York Times (1896).
19. Pigeon eggs: Laffan (1877) or faux eggs: Hammond (1918).
20. Harvest data for 1886: New York Times (1886); for 1904: New York Sun (1904).
21. Maryland industry declines: Truitt (1939).
22. Penned terrapins worth less: New York Times (1891).
23. Farming terrapins on Hog Island: Vallandigham (1894); in Crisfield: Baltimore Sun (1897), Wilson (1977).
24. LaVallette's farm thrives: Anonymous (1899), Meyer (2005), Fincham (2008); then collapses: Mayer (2005).

25. Not all farms were successful: New York Sun (1904).
26. Easton farm: Lowry (1888).
27. Protect terrapin beaches with netting: True (1887).
28. US Fish Commission consulted: Smith (1895b).
29. Terrapin study begun in Beaufort, NC: Coker (1906).
30. Study continued until 1949: Coker (1951), Wolfe (2000).
31. Remaining broodstock released in 1954: Wolfe (2000).
32. Maryland's 1878 conservation law: Velema and Speir (2007).
33. Laws not enforced, so modifications recommended: Thom (1898).
34. Terrapin excluders required: Lukacovic et al. (2002).
35. Demand declines: Carr (1952); perhaps due to a lack of alcohol: Woodhead (2007); or because the demand was just a fad: Carr (1952).

Chapter 6. Uncontrolled Market Hunting of Waterfowl

1. Eighty-seven species of waterfowl overwinter in Chesapeake Bay, and 138 species stop over while migrating south (autumn) or north (spring): Watts (2013). About one million waterfowl overwinter: www.fws.gov/chesapeakebay/migbird.html.
2. Wild celery: Lewis (1855), Hopper (1916).
3. Dimensions of Susquehanna flats: Davidson (1872).
4. Conservation and management efforts: Perry and Deller (1996).
5. Indians hunted wildfowl: Beverley (1722b); as did wealthy hunters: Sullivan (2003).
6. Drowned ducks become waterlogged and unsavory: Lewis (1851).
7. Duck prices: Davidson (1872).
8. Ducks frozen: Baltimore Gazette (1877).
9. Prices: New York Times (1888).
10. "Night guns": Grinnell (1901).
11. Sink box: Lewis (1851, 1855).
12. Ducks "unprecedentedly scarce": New York Times (1889).
13. Synopsis of Maryland's hunting laws from 1832 to 1936: Sullivan (2003).
14. Shooting times: Baltimore American (1887).
15. Response to Hornaday's survey: Hornaday (1898).
16. Educating children: Hegner (1907).
17. Influence of commercial hunters: Sullivan (2003).
18. Conservation concerns nationwide: Hornaday (1931).
19. Migratory Bird Treaty Convention passed: Hopper (1916).
20. Migratory Bird Treaty Act: www.law.cornell.edu/uscode/text/16/chapter-7/subchapter-II.

Chapter 7. Sturgeon

1. Indian legends and dances: www.native-languages.org/legends-sturgeon.htm.
2. Commercial sturgeon products: Mellinger (1904).
3. Coastal fisheries: Cobb (1900), Waldman (2013).
4. Decline in sturgeon numbers: Cobb (1900), Secor (2002), Waldman (2013).
5. Discarding valueless sturgeon: Smith (1914b).
6. Leaping sturgeon: De Voe (1867).
7. Sturgeon size: Elliot (1830), Martin (1835).
8. Hooking sturgeon: Elliot (1830).
9. Sturgeon sold for three cents per pound: Elliot (1830).
10. Food preservation review: Stevenson (1899b).
11. Roe sold to Europeans: Cobb (1900).
12. Sturgeon steaks: De Voe (1867).

13. "James River Bacon": Coleman (1892); "Charles City Bacon": Richmond Times-Dispatch (1904).
14. Scalding flesh: De Voe (1867).
15. Roe served to pigs: Tower (1908).
16. Congressmen and caviar: Tinkcom (1951).
17. Smaller gill nets used in 1884: Hovey (1884).
18. Mesh size of nets, and use of scows: Cobb (1900).
19. Yields nationwide and in Delaware Bay and decline in average catch per gill net: Tower (1908).
20. Causes for the decline: Tower (1908).
21. Water pollution degraded spawning grounds: Smith (1985).
22. Potomac harvests: Smith (1893).
23. Maryland and Virginia harvests: Smith (1895b); Tilp's (1978) history of a sturgeon fishery in Maryland includes details of boats and gear and their deployment, dressing the catch, and preparing caviar
24. Artisanal fishery in James River: Coleman (1892).
25. Reproduction data: www.nasps-sturgeon.org/sturgeons/biology/systematics/acipenser -oxyrinchus-oxyrinchus.aspx.

Chapter 8. Blue Crabs Hung On
This chapter is based upon Kennedy et al. (2007), which includes additional information about gear, regulations, processing practices, and blue crab fisheries nationally.
1. Virginia commercial harvests under-reported: Rhodes and Shabman (1994); recreational harvests: Cronin (1987).
2. Powhatan serves blue crabs: Hamor (1615); Tutor's meals included crabs: Fithian (1943); English traveler Parkinson reports: Wharton (1954).
3. Crabs as fish bait: Roberts (1905).
4. Crabbing was a casual operation: Rathbun (1887).
5. Crabbing rules: Roberts (1905).
6. Crabs shipped to nearby cities: Rathbun (1887).
7. Natural ice was a refrigerant: Donaldson and Nagengast (1994).
8. Bay fishery expands: Van Engel (1998).
9. Soft crabs pricier than hard crabs: Roberts (1905); Conservation Commission (1909).
10. Railroad to Crisfield: Warner (1976); Steamboat facilities: Smith (1891a), Roberts (1905), Wilson (1977), Johnson (1988).
11. Importance of Crisfield: Van Metre (1915); Deal Island: Roberts (1905).
12. Harvests in 1880: Rathbun (1887); in 1915: Churchill (1920); in mid-twentieth century: Van Engel (1962).
13. Soft crab culture began in New Jersey: Rathbun (1887).
14. Soft crab culture in Maryland: Smith (1891a); crabs first marketed outside Maryland: Roberts (1905).
15. Crabbing around Crisfield: Smith (1891a), New York Times (1894, 1895), Roberts (1905).
16. Numbers of soft crab vs. hard crab harvesters: Smith (1891a); Roberts (1905).
17. Prices: Smith (1891a).
18. Trays used: Smith (1891a), Roberts (1905), Churchill (1920).
19. Soft crabs to Canada: Roberts (1905).
20. Daily shipments from Crisfield: Conservation Commission (1909), Shell Fish Commission (1916).
21. Fried soft crabs for export: Baltimore Gazette (1877).
22. Peeler pounds: Roberts (1905).

23. New Jersey shedding floats: Rathbun (1887).
24. Crisfield floats: Smith (1891a).
25. Tawes' shedding systems: Warner (1976).
26. Closed circulation systems: Oesterling (1985). A closed system recirculates water within a series of tanks, with filtration units to maintain water quality, and with salinity, temperature, and oxygen controlled. Crabs can be shed away from the waterfront or from areas of poor water quality.
27. 1901 harvests: Roberts (1905); 1904 harvests: Conservation Commission (1909).
28. Trotlines: Roberts (1905), Shell Fish Commission (1916).
29. Hard crab packing: Roberts (1905), Shell Fish Commission (1916).
30. Popularity of deviled crab: New York Times (1895).
31. Crab meat details: Smith (1891a), Shell Fish Commission (1916).
32. Gasoline engines introduced: Roberts (1905).
33. Comparisons of a day's trotline catch: Brooks (1893a), Kennedy et al. (2007).
34. Periodic examinations of the fishery: Pearson (1942).
35. Churchill (1917).
36. Reports by Maryland agencies: Shell Fish Commission (1916), Conservation Commission (1923), Conservation Department (1927).
37. Virginia and Maryland laws on possessing sponge crabs: Churchill (1917).
38. Sette and Fiedler (1925).
39. Virginia bans any possession of sponge crabs: Conservation Department (1927).
40. Maryland begins collecting harvest data in 1928: Tarnowski (2000).
41. Virginia Blue Crab Sanctuary: Lambert et al. (2006).
42. Sanctuary protects female crabs: Lambert et al. (2006).
43. Difficulty in obtaining peeler crabs: Oesterling (1985).
44. Status of blue crab populations in 2017: www.chesapeakeprogress.com/abundant-life/blue-crab-abundance.

Chapter 9. Have Diminished Animal Abundances Remodeled the Bay's Food Webs?

1. Ecosystem subsidies: Flecker et al. (2010), Mattocks et al. (2017).
2. Feeding by shad: Walter and Olney (2003); by blueback herring: Creed (1985).
3. Changes in waterfowl diets: Perry and Deller (1996).
4. "Ghost fish": Waldman (2013).
5. Oysters filtering: Newell (1988).
6. Decline of oyster numbers: Wilberg et al. (2011).
7. Bay has shifted from a benthic to a pelagic system: Newell (1988).
8. Oysters and sea nettles: Breitburg and Fulford (2006).
9. Long-term ecological change: Klein and Thurstan (2016).

Afterword

1. Dynamic environments: Alagona et al. (2012), Klein and Thurstan (2016).
2. Great reversal in forestry: Ausubel (2008).
3. First marine reserve in New Zealand: www.newzealand.com/int/scenic-highlights+marine-reserves.
4. Spillover from reserves into fishing areas: http://ngm.nationalgeographic.com/2007/04/new-zealand-coast/warne-text.
5. Success of striped bass fishing moratorium: Richards and Rago (1999).
6. Oyster restoration: http://chesapeakebay.noaa.gov/oysters/oyster-restoration.
7. Spillover: Peters et al. (2017).

Appendix

1. Rau (1884) describes Indian fishing technologies, including those used in the Chesapeake Bay.
2. Comparative costs of nets: Milner (1876). The steam engine, lines, and net twine in the Stony Point seine fishery cost about $25,000 in 1875. Labor costs were additional. A pound net at the time cost $800 and could be fished by four or five people. A gill net was even cheaper and was fished by one to three people.
3. Shad float: Fitzgerald (1895), Gerstell (1988).
4. Fyke nets: True (1887), Brooks (1893a), Thom (1898), Coker (1949).
5. Crab scraping: Roberts (1905).
6. Lay of trotline: Brewington (1953).

References

Alagona, P.S., J. Sandlos, and Y.F. Wiersma. 2012. Past imperfect: Using historical ecology and baseline data for conservation and restoration projects in North America. Environmental Philosophy 9:49–70.

Alford, J.J. 1975. The Chesapeake oyster fishery. Annals of the Association of American Geographers 65:229–239.

Alleway, H.K., and S.D. Connell. 2015. Loss of an ecological baseline through the eradication of oyster reefs from coastal ecosystems and human memory. Conservation Biology 29:795–804.

Alleway, H.K., R.H. Thurstan, P.R. Lauer, and S.D. Connell. 2016. Incorporating historical data into aquaculture planning. ICES Journal of Marine Science 73:1427–1436.

Alsop, G. 1666. A Character of the Province of Maryland. London. 111 pages. Reprinted in 1902 with an Introduction and Notes by N.D. Mereness. Burrows Brothers Company, Cleveland, OH.

Andrews, J.D. 1979. Oyster diseases in Chesapeake Bay. Marine Fisheries Review 41(1–2): 45–53.

Annual Report of the Fish Commissioners of the State of Virginia for the Year 1875. R.F. Walker, Richmond. 34 pages.

Anonymous. 1869. Oysters of the Chesapeake—Their propagation and culture. Pages 341–347 in Report of the Commissioner of Agriculture for the Year 1868. Washington, DC.

———. 1896. Salmon, ducks, and terrapin. Forest and Stream 46(20):389.

———. 1899. Terrapin farming. Forest and Stream 53:191.

Ausubel, J.H. 2008. Future knowledge of life in oceans past. Pages xix–xxvi in D.J. Starkey, P. Holm, and M. Barnard (eds.). 2008. Oceans Past. Management Insights from the History of Marine Animal Populations. Earthscan, Sterling, VA. 223 pages.

Baird, S.F. 1873. Description of apparatus used in capturing fish on the sea-coast and lakes of the United States. Pages 253–274 in Report on the Condition of the Sea Fisheries of the South Coast of New England in 1871 and 1872. United States Commission of Fish and Fisheries, Part I. Washington, DC.

———. 1889. The sea fisheries of eastern North America. Pages 3–224 in Report of the Commissioner for 1886. US Commission of Fish and Fisheries, Part 14. Washington, DC.

Baltimore American. 1887. Canvasback ducks. Thousands killed on the Susquehanna River—how slaughtered. Reprinted in the New York Times, November 6.

Baltimore Gazette. 1877. Fish and Fowl. How oysters, terrapin, fish, and game are shipped. Reprinted in the New York Times, February 17.

Baltimore Sun. 1887. Maryland terrapins. How they are caught and cooked—tricks of caterers. Reprinted in the New York Times, December 12.

———. 1893. Editorial entitled "An Economic Question of Importance." February 18.

———. 1894. How terrapin prices vary. Reprinted in the New York Times, February 25.

———. 1897. Terrapin culture. A Maryland expert talks entertainingly about it—a profitable industry. Reprinted in the New York Times, December 26.

———. 1903. Editorial entitled "Good Roads and Oyster Planting." July 31.

———. 1905. Editorial entitled "The Oyster and the Politician." June 17.

Bay Journal. 2007. MD legislators ban commercial harvest of diamondback terrapins. April 7. www.bayjournal.com/article/md_legislators_ban_commercial_harvest_of_diamondback _terrapins.

Baylor, J.B. 1895. Survey of Oyster Grounds in Virginia. J.H. O'Bannon, Superintendent of Public Printing, Richmond, VA. 16 pages.

Beverley, R. 1722a. The History of Virginia, in Four Parts. Book II. The Natural Production and Conveniences of the Country, Suited to Trade and Improvement. Chapter V. Of the Fish. Reprinted in 1855 from the author's second revised edition, London, 1722. J.W. Randolph, Richmond, VA. Pages 117–122.

———. 1722b. The History of Virginia, in Four Parts. Book II. The Natural Production and Conveniences of the Country, Suited to Trade and Improvement. Chapter VI. Of Wild Fowl and Hunted Game. Reprinted in 1855 from the author's second revised edition, London, 1722. J.W. Randolph, Richmond, VA. Pages 123–126.

Boesch, D.F. 2006. Scientific requirements for ecosystem-based management in the restoration of Chesapeake Bay and Coastal Louisiana. Ecological Engineering 26:6–26.

Boesch, D.F., R.B. Brinsfield, and R.E. Magnien. 2001. Chesapeake Bay eutrophication: Scientific understanding, ecosystem restoration and challenges for agriculture. Journal of Environmental Quality 30:303–320.

Bowers, G.M. 1907. Statistics of the Fisheries of the Middle Atlantic States for 1904. Bureau of Fisheries Document No. 609. Washington, DC. 121 pages.

Bowman, I. (Chairman). 1948. Report of the Commission on Conservation of Natural Resources to the Governor of Maryland, Annapolis. 91 pages.

Boynton, W.R., J.H. Garber, R. Summers, and W.M. Kemp. 1995. Inputs, transformations, and transport of nitrogen and phosphorus in Chesapeake Bay and selected tributaries. Estuaries 18(1B):285–314.

Breitburg, D.L., and R.S. Fulford. 2006. Oyster–sea nettle interdependence and altered control within the Chesapeake Bay ecosystem. Estuaries and Coasts 29:776–784.

Brewington, M.V. 1953. Chesapeake Bay. A Pictorial Maritime History. Bonanza Books, New York. 238 pages.

Brooks, W.K. 1891. The Oyster. A Popular Summary of a Scientific Study. Johns Hopkins Press, Baltimore, MD. 240 pages.

———. 1893a. Chapter VII. Fish and Fisheries. Pages 239–263 in (No editor). Maryland. Its Resources, Industries and Institutions. Board of World's Fair Managers of Maryland. Baltimore, MD.

——— 1893b. Chapter VIII. The Oyster. Pages 264–312 in (No editor). Maryland. Its Resources, Industries and Institutions. Board of World's Fair Managers of Maryland. Baltimore, MD.

———. 1905. The Oyster. A Popular Summary of a Scientific Study. Second and revised edition. Johns Hopkins Press, Baltimore, MD. 240 pages.

———. 1996. The Oyster. A Popular Summary of a Scientific Study, with an Introduction by Kennedy T. Paynter, Jr. Johns Hopkins University Press, Baltimore, MD. 246 pages.

Brooks, W.K., J.I. Waddell, and W.H. Legg. 1884. Report of the Oyster Commission of the State of Maryland. James Young, Annapolis, MD. 183 pages, 13 plates, and 6 maps.

Brown, A.C. 1940. The Old Bay Line 1840–1940. Bonanza Books, New York. 176 pages.

Brush, G. 2017. Decoding the Deep Sediments. The Ecological History of Chesapeake Bay. Maryland Sea Grant Publication UM-SG-CP-2017-01. College Park, MD. 63 pages.

Burk, J. 1805. The History of Virginia, from Its First Settlement to the Present Day, Volume II. Dickson & Pescud, Petersburg, VA. 335 pages + 52-page appendix.

Burnaby, A. 1775. Travels through the Middle Settlements in North America in the Years 1759 and 1760. London. Reprinted as R.R. Wilson. 1904. Burnaby's Travels through North America. Reprinted from the 3rd Edition of 1798 with Introduction and Notes. A. Wessels Company, New York. 265 pages.

Byrd, W. 1728. William Byrd's Natural History of Virginia; or, The Newly Discovered Eden. Edited and translated in 1940 from a German version of 1737 by R.C. Beatty and W.J. Mulloy. Dietz Press, Richmond, VA. 204 pages.

Cargo, D.G. 1954. Maryland's Commercial Fishing Gears. III. The Crab Gears. Maryland Board of Natural Resources, Educational Series 36. Solomons, MD. 18 pages.

Carr, A. 1952. Handbook of Turtles. Cornell University Press, Ithaca, NY. 542 pages.

Chesapeake Bay Fisheries Ecosystem Advisory Panel. 2006. Fisheries Ecosystem Planning for Chesapeake Bay. American Fisheries Society, Trends in Fisheries Science and Management 3, Bethesda, MD. 410 pages.

Chesapeake Bay Program. 2012. Population. www.chesapeakebay.net/indicators/indicator /chesapeake_bay_watershed_population.

Christy, F.T. 1964. The exploitation of a common property natural resource: The Maryland oyster industry. PhD Dissertation. University of Michigan, Ann Arbor. 444 pages.

Churchill, E.P., Jr. 1917. The Conservation of the Blue Crab of Chesapeake Bay. Unpublished manuscript available from Chesapeake Biological Laboratory, Solomons, MD. 19 pages.

———. 1920. Crab industry of Chesapeake Bay. Pages 1–25 in Report of the US Commissioner of Fisheries for 1918, Appendix 4. Washington, DC.

Cobb, J.N. 1900. The sturgeon fishery of Delaware River and Bay. Report of the Commissioner for the Year ending June 30, 1899, Volume 25:369–380 + 3 plates. US Commission of Fish and Fisheries, Washington, DC.

Coker, C.M. 1949. Maryland's Commercial Fishing Gears. I. The Fin-Fish Gears. Maryland Board of Natural Resources, Educational Series 18. Solomons, MD. 37 pages.

Coker, R.E. 1906. The natural history and cultivation of the diamond-back terrapin with notes on other forms of turtles. North Carolina Geological Survey Bulletin 14:1–67.

———. 1951. The diamond-back terrapin in North Carolina. Pages 219–230 in H.F. Taylor (ed.). Survey of Marine Fisheries of North Carolina. University of North Carolina Press, Chapel Hill, NC.

Coleman, C.W. 1892. Sturgeon fishing in the James. The Cosmopolitan 13(3):366–373.

Collins, J.H. 1924. The Story of Canned Foods. E.P. Dutton, New York. 251 pages.

Collins, J.W. 1891. Statistical review of the coast fisheries of the United States. Pages 271–378 in Report of the Commissioner of Fish and Fisheries for 1888. Washington, DC.

Conservation Commission. 1909. Pages 155–179 in Report of the Conservation Commission of Maryland for 1908–1909. Baltimore, MD.

———. 1918. Second Annual Report of the Conservation Commission of Maryland 1917. Baltimore, MD. 80 pages.

———. 1923. Seventh Annual Report of the Conservation Commission of Maryland 1922. Baltimore, MD. 87 pages.

Conservation Department. 1927. Fourth Annual Report of the Conservation Department of the State of Maryland 1926. Baltimore, MD. 131 pages.

Cooper, S.R. 1995. Chesapeake Bay watershed historical land use: Impact of water quality and diatom communities. Ecological Applications 5:703–723.

Cooper, S.R., and G.S. Brush. 1993. A 2,500-year history of anoxia and eutrophication in Chesapeake Bay. Estuaries 16:617–626.

Creed, R.P., Jr. 1985. Feeding, diet, and repeat spawning of blueback herring, *Alosa aestivalis*, from the Chowan River, North Carolina. Fishery Bulletin 83:711–716.

Cronin, L.E. 1950. The Maryland Crab Industry, 1949. Maryland Board of Natural Resources, Publication 84. Solomons, MD. 41 pages.

———— (ed.). 1987. Report of the Chesapeake Bay blue crab management workshop, Waldorf, Maryland. November 9–10, 1987. No publisher named. 68 pages.

Curtin, P.D., G.S. Brush, and G.W. Fisher. 2001. Discovering the Chesapeake. The History of an Ecosystem. Johns Hopkins University Press, Baltimore, MD. 385 pages.

Danckaerts, J. 1679–1680. Journal of Jasper Danckaerts, 1679–1680. Edited in 1913 by B.B. James and J.F. Jameson. Charles Scribner's Sons, New York. 310 pages.

Davidson, H. 1872. Report on the Oyster Fisheries: Potomac River Shad and Herring Fisheries, and the Water-Fowl of Maryland to His Excellency the Governor and Other Commissioners of the State O.P. Force. S.S. Mills, L.F. Colton & Co., Annapolis, MD. 48 pages.

De Voe, T.F. 1867. The Market Assistant, Containing a Brief Description of Every Article of Human Food Sold in the Public Markets of New York, Boston, Philadelphia, and Brooklyn. Hurd and Houghton, New York. 455 pages.

Dickens, C. 1837. The Posthumous Papers of the Pickwick Club. Chapman and Hall, London. 609 pages.

Dilts, J.D. 1993. The Great Road: The Building of the Baltimore and Ohio, the Nation's First Railroad, 1828–1853. Stanford University Press, Stanford, CA. 472 pages.

Dodge, R.I. 1877. The Plains of the Great West. G.P. Putnam's Sons, New York. 452 pages.

Donaldson, B., and B. Nagengast. 1994. Heat and Cold: Mastering the Great Indoors. American Society of Heating, Refrigerating and Air-Conditioning Engineers, Atlanta, GA. 339 pages.

Ducatel, J.T. 1837. Outline of the physical geography of Maryland, embracing its prominent geological features. Transactions of the Maryland Academy of Science and Literature 1:24–54.

Durand of Dauphine. 1686. A Frenchman in Virginia: Being the Memoirs of a Huguenot Refugee in 1686. Translated in 1923 by a Virginian (Fairfax Harrison). Privately printed. Richmond, VA. 146 pages.

Earle, S. 1932. The fisheries of Chesapeake Bay. Transactions of the American Fisheries Society 62:43–49.

Elliot, J. 1830. Historical Sketches of the Ten Miles Square Forming the District of Columbia. J. Elliot, Jr., Washington, DC. 554 pages.

Farenthold, D.A. 2006. A Dead Indian Language Is Brought Back to Life. Washington Post, December 12, page A1.

Fincham, M.W. 2008. The men who would be kings. Chesapeake Quarterly 7(4):10–14.

Fisheries Statistics Division. 1990. Historical Catch Statistics, Atlantic and Gulf Coast States, 1879–1989. Current Fishery Statistics No. 9010, Historical Series Nos. 5–9 Revised. NTIS No. PB-93-174274. National Technical Information Service, Alexandria, VA.

Fithian, P. 1943. Journals and Letters, 1773–1774. Edited by H.D. Farish. Williamsburg, VA. Not seen; quoted in Wharton 1957.

Fitzgerald, D.B. 1894. Duck shooting in Maryland. The Cosmopolitan 18(November):61–68.

————. 1895. On a shad-float. Lippincott's Monthly Magazine 55(May):692–696.

Flecker, A.S., P.B. McIntyre, J.W. Moore, J.T. Anderson, B.W. Taylor, and R.O. Hall, Jr. 2010. Migratory fishes as material and process subsidies in riverine ecosystems. American Fisheries Society Symposium 73:559–592.

Fleet, H. 1632. A Brief Journal of a Voyage made on the Bark "Warwick" and other Parts of the Continent of America. Not seen but reported by E.D. Neill. 1871. The English Colonization of America during the Seventeenth Century, pages 221–237. Strahan & Co., London.

Flemer, D.A., G.B. Mackiernan, W. Nehlsen, V.K. Tippie, R.B. Biggs, D. Blaylock, N.H. Burger, L.C. Davidson, D. Haberman, K.S. Price, and J.L. Taft. 1983. Chesapeake Bay: A Profile of Environmental Change. Chesapeake Bay Program, Annapolis, MD. 200 pages.

Forbush, E.H. 1912. A History of the Game Birds, Wild-Fowl and Shore Birds of Massachusetts and Adjacent States. Massachusetts State Board of Agriculture, Boston. 622 pages.

Fowler, G. 1881. Statement of Gilbert Fowler, age 89, reported in H. Wright. 1884. On the early shad fisheries of the North Branch of the Susquehanna River. Pages 619–642 in Report of the Commissioner for 1881, Part 9. US Commission of Fish and Fisheries, Washington, DC.

Frey, J. 1893. Reminiscences of Baltimore. Maryland Book Concern, Baltimore, MD. 468 pages.

Gallatin, A. 1808. Report of the Secretary of the Treasury, on the Subject of Public Roads and Canals; Made in Pursuance of a Resolution of the Senate, of March 2, 1807. R.C. Weightman, Washington, DC. 94 pages.

Gallivan, M. 2012. Native history in the Chesapeake: The Powhatan chiefdom and beyond. Pages 310–322 in T.R. Pauketat (ed.). The Oxford Handbook of North American Archaeology. Oxford, UK.

Gay, J. 1892. The shad streams of Pennsylvania. Pages 151–187 in Report of the State Commissioners of Fisheries for the Years 1889–90–91. Harrisburg, PA.

Gerstell, R. 1998. American Shad in the Susquehanna River Basin: A Three-Hundred-Year History. The Pennsylvania State University Press, University Park, PA. 217 pages.

Glover, T. 1676. An account of Virginia, its scituation [sic], temperature, productions, inhabitants and their manner of planting and ordering tobacco &c. Philosophical Transactions of the Royal Society, June 20, 1676. Reprinted 1904 by B.H. Blackwell, Oxford.

Goode, G.B. 1883a. The first decade of the United States Fish Commission: Its plan of work and accomplished results, scientific and economical. Pages 53–62 in Report of the Commissioner for 1880, II. Appendix to Report of Commissioner, Appendix A, General, II. Washington, DC.

———. 1883b. A Review of the Fishery Industries of the United States and the work of the US Fish Commission. William Clowes and Sons Limited, London. 86 pages.

——— (ed.). 1887. The Fisheries and Fishery Industries of the United States. Section II. A Geographical Review of the Fisheries Industries and Fishing Communities for the Year 1880. US Commission of Fish and Fisheries, Washington, DC. 787 pages.

Grave, C. 1912. A manual of oyster culture in Maryland. Pages 279–349 in Fourth Report of the Board of Shell Fish Commissioners. Baltimore, MD.

Green, B.K., F.S. Revell, and W.H. Maltbie (eds.). 1916. Seventh Report of the Shell Fish Commission 1914 and 1915. Baltimore, MD. 78 pages.

Grinnell, G.B. 1901. American Duck Shooting. Willis McDonald & Company, New York. 623 pages.

Haman, H.B. 1893. Oysters and roads. Address delivered before the Maryland Convention for Good Roads, Baltimore, January 12, 1893. Maryland Road League, Baltimore, MD. 24 pages with charts.

Hammer, R.C. 1948. Present status of the Chesapeake Bay oyster bars in Maryland. Proceedings of the National Shellfisheries Association 1947:8–10.

Hammond, J. 1656. Leah and Rachel, or, the Two Fruitfull [sic] Sisters Virginia and Maryland: Their Present Condition, Impartially Stated and Related. T. Mabb, London. 24 pages.

Hammond, M.M.E. 1918. The Swedish, French, and American Cook Book. Privately printed, New York. 480 pages.

Hamor, R. 1615. A True Discourse of the Present Estate of Virginia. Reprinted in E.W. Haile (ed.). 1998. Jamestown Narratives. Eyewitness Accounts of the Virginia Colony. The First Decade: 1607–1617, pages 792–856. RoundHouse, Champlain, VA.

Handley, B.M. 2001. The Blue Goose Midden (8IR15): A Malabar II occupation on the Indian River Lagoon. The Florida Anthropologist 54(3–4):103–121.

Hargis, W.J., Jr. 1999. The evolution of the Chesapeake oyster reef system during the Holocene Epoch. Pages 5–23 in M.W. Luckenbach, R. Mann, and J.A. Wesson (eds.). Oyster Reef Habitat Restoration: A Synopsis and Synthesis of Approaches. VIMS Press, Gloucester Point, VA.

Hariot, T. 1590. A Brief and True Report of the New Found Land of Virginia. Reprinted as Hariot's Narrative of the First Plantation of Virginia in 1585. Reprinted in 1893 from the Edition of 1590 with De Bry's Engravings. Bernard Quaritch, London. 111 pages. http://docsouth.unc.edu/nc/hariot/hariot.html.

Haven, D.S., W.J. Hargis, Jr., and P.C. Kendall. 1978. The Oyster Industry of Virginia: Its Status, Problems, and Promise. Special Papers in Marine Science No. 4. Virginia Institute of Marine Science, Gloucester Point, VA. 1024 pages.

Hegner, R.W. 1907. Nature-studies with birds for the elementary school. The Elementary School Teacher 7(6):348–354.

Hildebrand, S.F., and W.C. Schroeder. 1928. Fishes of Chesapeake Bay. Bulletin US Bureau of Fisheries 43(1):1–366.

Hobbs, C.H., III, J.P. Halka, R.T. Kerhin, and M.J. Carron. 1992. Chesapeake Bay sediment budget. Journal of Coastal Research 8:292–300.

Holberton, W. 1892. Shad-fishing on the Susquehanna River. Harper's Weekly 36(1843):372–373.

Holmes, O.W. 1860. The Professor at the Breakfast-Table: With the Story of Iris. Ticknor and Fields, Boston. 410 pages.

Hopper, G.L. 1916. Ducking on the Susquehanna flats, past and present. Pages 27–34 in W.C. Hazelton (compiler). Tales of Duck and Goose Shooting. Press of Eastman Bros., Chicago, IL.

Hornaday, W.T. 1898. The destruction of our birds and mammals: A report on the results of an inquiry. Pages 77–126 in Second Annual Report of the New York Zoological Society. New York.

———. 1931. Thirty Years War for Wildlife: Gains and Losses in the Thankless Task. Charles Scribner's Sons, New York. 316 pages.

Hovey, H.C. 1884. The sturgeon fishery. Bulletin US Fish Commission for 1884, 4:346–348.

Howard, G.W. 1873. Oyster, Fruit and Vegetable Packing. Pages 101–104 in The Monumental City. Its Past History and Present Resources. J.D. Elders & Co., Baltimore, MD.

Hungerford, J. 1859. The Old Plantation and What I Gathered There in an Autumn Month. Harper and Brothers, New York. 369 pages.

Ingersoll, E. 1881. The History and Present Condition of the Fishery Industries: The Oyster Industry. US Census Bureau, 10th Census. Department of the Interior, Washington, DC. 251 pages.

———. 1887. The oyster industry. Pages 507–565 in The Fisheries and Fishery Industries of the United States, Section V. History and Methods of the Fisheries, Volume II. US Government Printing Office, Washington, DC.

Johnson, E.R. 1908. Geographic influences affecting the early development of American commerce. Bulletin of the American Geographical Society 40:129–143.

Johnson, P.J. (ed.). 1988. Working the Water. The Commercial Fisheries of Maryland's Patuxent River. Calvert Marine Museum, Solomons, MD, and University Press of Virginia, Charlottesville, VA. 218 pages.

Jones, C.G., J.H. Lawton, and M. Shachak. 1994. Organisms as ecosystem engineers. Oikos 69:373–386.

Judge, E.S. 1895. American canning interests. Pages 395–400 in C.M. Depew (ed.). One Hundred Years of American Commerce, Volume 2. D.O. Haynes & Co., New York.

Kalm, P. 1772. Travels into North America, 2nd Edition. Volume 2. T. Lowndes, London. 433 pages.

Kee, E. 2006. Saving Our Harvest: The Story of the Mid-Atlantic Region's Canning and Freezing Industry. CTI Publications, Baltimore, MD. 436 pages.

Keiner, C. 2009. The Oyster Question. The University of Georgia Press, Athens, GA. 331 pages.

Kemp, W.M., and W.R. Boynton. 1992. Benthic-pelagic interactions: Nutrients and oxygen dynamics. Pages 149–221 in D.E. Smith, M. Leffler, and G. Mackiernan (eds.). Oxygen Dynamics in the Chesapeake Bay: A Synthesis of Recent Research. Maryland Sea Grant College Program, College Park, MD.

Kemp, W.M., W.R. Boynton, J.E. Adolf, D.F. Boesch, W.C. Boicourt, G. Brush, J.C. Cornwell, T.R. Fisher, P.M. Glibert, J.D. Hagy, L.W. Harding, E.D. Houde, D.G. Kimmel, W.D. Miller, R.I.E. Newell, M.R. Roman, E.M. Smith, and J.C. Stevenson. 2005. Eutrophication of Chesapeake Bay: Historical trends and ecological interactions. Marine Ecology Progress Series 303:1–29.

Kemp, W.M., J. Faganeli, S. Puskaric, E.M. Smith, and W.R. Boynton. 1999. Pelagic-benthic coupling and nutrient cycling. Pages 295–339 in T.C. Malone, A. Malej, L.W. Harding, Jr., N. Smodlaka, and R.E. Turner (eds.). Ecosystems at the Land-Sea Margin: Drainage Basin to Coastal Sea. Coastal and Estuarine Studies Volume 55. American Geophysical Union, Washington, DC.

Kemp, W.T., W.H. Killian, and J.E. White. 1918. Second Annual Report of the Conservation Commission of Maryland 1917. Baltimore, MD.

Kennedy, V.S. 1989. The Chesapeake Bay oyster fishery: Traditional management practices. Pages 455–477 in John F. Caddy (ed.). Marine Invertebrate Fisheries: Their Assessment and Management. John Wiley and Sons, New York.

———. 2018. History of commercial fisheries and artificial propagation. Chapter 13 in W.M. Roosenburg and V.S. Kennedy (eds.). Ecology and Conservation of the Diamond-backed Terrapin. Johns Hopkins University Press, Baltimore, MD.

Kennedy, V.S., and L.L. Breisch. 1983. Sixteen decades of political management of the oyster fishery in Maryland's Chesapeake Bay. Journal of Environmental Management 16:153–171.

Kennedy, V.S., and K. Mountford. 2001. Human influences on aquatic resources in the Chesapeake Bay watershed. Pages 191–219 in P.D. Curtin, G. Brush, and G.W. Fisher (eds.). Discovering the Chesapeake. Johns Hopkins University Press, Baltimore, MD.

Kennedy, V.S., M. Oesterling, and W.A. Van Engel. 2007. History of blue crab fisheries on the US Atlantic and Gulf Coasts. Pages 655–709 in V.S. Kennedy and L.E. Cronin (eds.). The Blue Crab Callinectes sapidus. Maryland Sea Grant, College Park, MD.

King, E. 1875. Baltimore: The Liverpool of America. Scribner's Monthly 9(6):681–695.

Klapp, H.M. (ed.). 1853. Krider's Sporting Anecdotes. A. Hart, Philadelphia. 286 pages.

Klein, E.S., and R.H. Thurstan. 2016. Acknowledging long-term ecological change: The problem of shifting baselines. Pages 11–29 in K. Schwerdtner Máñez and B. Poulsen (eds.). Perspectives on Oceans Past. Springer, Dordrecht.

Kobell, R. 2017. Bay's oyster harvest numbers closing in on wild fishery. Chesapeake Bay Journal 27(8):9–10.

Laffan, W.M. 1877. Canvas-back and terrapin. Scribner's Monthly 15(1):1–13.

Lambert, D.M., R.M. Lipcius, and J.M. Hoenig. 2006. Assessing effectiveness of the blue crab spawning stock sanctuary in Chesapeake Bay using tag-return methodology. Marine Ecology Progress Series 321:215–225.

Lang, V. 1961. Follow the Water. John F. Blair, Winston-Salem, NC. 222 pages.

Leffingwell, W.B. 1890. The canvas-back duck. Pages 403–420 in W.B. Leffingwell (ed.). Shooting on Upland, Marsh and Stream. Rand, McNally & Company, New York.

Lewis, E.J. 1851. Hints to Sportsmen, Containing Notes on Shooting. Lea and Blanchard, Philadelphia. 366 pages.

———. 1855. The American Sportsman, 2nd edition. Lippincott, Grambo and Co., Philadelphia. 319 pages.

Limburg, K.E., and J.R. Waldman. 2009. Dramatic declines in North American diadromous fishes. BioScience 59:955–965.

Livingood, J.W. 1941. The canalization of the lower Susquehanna. Pennsylvania History 8(2):131–147.

Longfellow, H.W. 1855. The Song of Hiawatha. David Bogue, London. 316 pages.

Lowry, R.C. 1888. Terrapin culture (letter to the editor). Forest and Stream. December 13, 1888:31. http://eshore.vcdh.virginia.edu/node/1889.

Lukacovic, R., J. Uphoff and H. Speir. 2002. An evaluation of the status of the diamondback terrapin *Malaclemys terrapin* in Maryland's coastal bays and the effects of crab pot by-catch on the population. Maryland Department of Natural Resources Fisheries Technical Memo No. 27:1–18.

MacKenzie, C.L., Jr. 1996. History of oystering in the United States and Canada, featuring the eight greatest oyster estuaries. Marine Fisheries Review 58(4):1–78.

Malone, T.C., A. Malej, L.W. Harding, Jr., N. Smodlaka, and R.E. Turner (eds.). 1999. Ecosystems at the Land-Sea Margin: Drainage Basin to Coastal Sea. Coastal and Estuarine Studies Volume 55. American Geophysical Union, Washington, DC. 381 pages.

Maltbie, W.H. 1914. Statements submitted by the Board of Shell Fish Commissioners to the State of Maryland. Available in Pratt Library, Baltimore, MD. Call No. X3H 465 M3 A473.

Martin, J. (Publisher). 1835. Fisheries. Pages 480–481 in A New and Comprehensive Gazetter of Virginia and the District of Columbia. Moseley & Tomkins, Charlottesville, VA.

Maryland Commissioners Pan-American Exposition. 1901. Water products. Pages 30–34 in Maryland and Its Natural Resources. Maryland Geological Survey, Baltimore, MD.

Massmann, W.H. 1961. A Potomac River shad fishery, 1814–1824. Chesapeake Science 2:76–81.

Mattocks, S., C.H. Hall, and A. Jordaan. 2017. Damming, lost connectivity, and the historical role of anadromous fish in freshwater ecosystem dynamics. BioScience 67:713–728.

Mayer, B. 1871. Baltimore: Past and Present. With Biographical Sketches of its Representative Men. Richardson & Bennett, Baltimore, MD. 562 pages.

McCauley, R.H., Jr. 1945. The Reptiles of Maryland and the District of Columbia. Privately published, Hagerstown, MD. 194 pages.

McDonald, M. 1887. The fisheries of Chesapeake Bay and its tributaries. Pages 637–654 in G.B. Goode (ed.). The Fisheries and Fishery Industries of the United States. Section V. Volume I. Part XII, 3. US Government Printing Office, Washington, DC.

McPhee, J. 2002. The Founding Fish. Farrar, Straus, and Giroux, New York. 358 pages.

Meehan, W.E. 1893. Fish, fishing and fisheries of Pennsylvania. Pennsylvania's Fish Exhibit at the World's Columbian Exposition, Chicago, 1893. Compiled for the State Fish Commissioners. E. K. Meyers, Harrisburg, PA. 106 pages.

———. 1898. The herring industry. Page 38 in Report of the State Commissioners of Fisheries for the Year 1898. Wm. Stanley Ray, State Printer, Harrisburg, PA.

Mellinger, M.J. (Compiler). 1904. Sturgeon. Pennsylvania's Fish Cultural Work and Exhibit at World's Fair, St. Louis, 1904. Commonwealth of Pennsylvania, Department of Fisheries Bulletin 2:39–41.

Mencken, H.L. 1940. Happy Days: 1880–1892. Alfred A. Knopf, New York. 313 pages.

Meyer, E.L. 2005. Easy come, easy go. Chesapeake Bay Magazine 34 (April 2005):74–79, 104–107.

Michel, F.L. 1702. Report of the journey of Francis Louis Michel from Berne, Switzerland, to Virginia, October 2, 1701–December 1, 1702. Translated from the French and edited by

William J. Hinke in 1916. The Virginia Magazine of History and Biography 24(1):1–43, 113–141, 275–303.

Milner, J.W. 1876. Report of the Triana trip. Pages 351–362 in Report of the Commissioner for 1873–4 and 1874–5. Part III. Appendix B (XVIII). US Commissioner of Fish and Fisheries, Washington, DC.

Mitchell, W.J., C. Grave, and B.K. Green. 1907. First Report of the Shell Fish Commission of Maryland. Sun Job Printing Office, Baltimore, MD. 231 pages.

Moore, H.F. 1910. Condition and extent of the oyster beds of the James River, Virginia. Bureau of Fisheries Document No. 729:7–83.

Moseley, A. 1877. Annual Report of the Fish Commissioner of the State of Virginia for the Year 1877. R.F. Walker, Superintendent Public Printing, Richmond, VA. 58 pages.

National Marine Fisheries Service (NMFS). 2016. Fisheries Economics of the United States, 2014. US Dept. of Commerce, NOAA Tech. Memo. NMFS-F/SPO-163. 237 pages. www.st.nmfs.noaa.gov/economics/publications/feus/fisheries_economics_2014/index.

New York Sun. 1904. How to save the terrapin. They are decreasing so fast that they won't last long. January 3.

New York Times. 1886. The terrapin industry. How the palatable reptile is captured. Its favorite haunts in Chesapeake Bay and how terrapin farming is carried on. November 21.

———. 1888. Canvas-back duck trust. No monopoly in terrapins possible. February 5.

———. 1889. Scarcity of the canvas-backs and fears for their extinction. November 24.

———. 1891. Our tables will suffer. Terrapin and canvas-back surely disappearing. The oyster will follow them unless measures are taken for its preservation. What some of the States have done. February 15.

———. 1894. Crabs and their catching. Important and lucrative industry of Chesapeake Bay. July 1.

———. 1895. Many millions of crabs. Crisfield, on the Chesapeake Bay, is where they are caught. May 12.

———. 1896. An epicure and a terrapin. Enthusiastic over a delicacy now growing scarce. Many persons deceived with common turtles—the proper way to cook the real thing. January 12.

Newell, R.I.E. 1988. Ecological changes in Chesapeake Bay: Are they the result of overharvesting the Eastern oyster (*Crassostrea virginica*)? Pages 536–546 in M.P. Lynch and E.C. Krome (eds.). Understanding the Estuary: Advances in Chesapeake Bay Research. Chesapeake Research Consortium Publication 129 (CBP/TRS 24/88). Gloucester Point, VA.

Nichol, A.J. 1937. The oyster-packing industry of Baltimore: Its history and current problems. Bulletin, Chesapeake Biological Laboratory. Solomons Island, MD. 32 pages.

Nichols, T.L. 1864. Forty Years of American Life, Volume 1. John Maxwell and Company, London. 408 pages.

Odum, W.E. 1970. Insidious alteration of the estuarine environment. Transactions of the American Fisheries Society 99:836–847.

Oesterling, M.J. 1985. The Virginia soft crab fishery and assistance initiatives. Pages 89–90 in H.M. Perry and R.F. Malone (eds.). Proceedings of the National Symposium on the Soft-Shelled Blue Crab Fishery, February 12–13, 1985. Gulf Coast Research Laboratory, Biloxi, MS.

Officer, C.B., R.B. Biggs, J.L. Taft, L.E. Cronin, M.A. Tyler, and W.R. Boynton. 1984. Chesapeake Bay anoxia: Origin, development, and significance. Science 223:22–27.

Page, J.R. 1877. Oyster-beds and oyster-culture in Virginia. Pages 23–25 in Appendix No. 1, Annual Report of the Fish Commissioner of the State of Virginia for the Year 1877. R.F. Walker, Superintendent Public Printing, Richmond, VA.

Parsons, J.S. 1917. The fisheries of Chesapeake Bay and the co-operation of Maryland and Virginia for their conservation. Pages 18–24 in 18th Annual Report of the Commission of Fisheries of Virginia. Davis Bottom, Richmond, VA.

Pauly, D. 1995. Anecdotes and the shifting baseline syndrome of fisheries. Trends in Ecology and Evolution 10:430.

Pearson, J.C. 1942. Decline in abundance of the blue crab, *Callinectes sapidus*, in Chesapeake Bay during 1940, and 1941, with suggested conservation measures. US Fish and Wildlife Service Special Scientific Report 16. 27 pages.

Perez, C. 2007. Great surges of development and alternative forms of globalization. Working Papers in Technology Governance and Economic Dynamics No. 15. The Other Canon Foundation, Norway. 30 pages.

Perley, M.H. 1850. Report on the sea and river fisheries of New Brunswick, within the Gulf of Saint Lawrence and Bay of Chaleur. J. Simpson, Fredericton, New Brunswick. 137 pages.

Perry, M.C., and A.S. Deller. 1996. Review of factors affecting the distribution and abundance of waterfowl in shallow-water habitats of Chesapeake Bay. Estuaries 19(2A):272–278.

Peters, J.W., D.B. Eggleston, B.J. Puckett, and S.J. Theuerkauf. 2017. Oyster demographics in harvested reefs vs. no-take reserves: Implications for larval spillover and restoration success. Frontiers in Marine Science 27. https://doi.org/10.3389/fmars.2017.00326.

Porter, W.T. 1846. Instructions to Young Sportsmen, in all that relates to Guns and Shooting, by Lieut. Col. P. Hawker, to which is added the Hunting and Shooting of North America, with Descriptions of the Animals and the Birds. First American edition from the ninth London Edition. Lea and Blanchard, Philadelphia. 458 pages.

Potter, S.R. 1993. Commoners, Tribute, and Chiefs: The Development of Algonquian Culture in the Potomac Valley. University Press of Virginia, Charlottesville. 267 pages.

Power, G. 1970. More about oysters than you wanted to know. Maryland Law Review 30:199–225.

Radcliffe, L. 1922. Fishery Industries of the United States: Report of the Division of Fishery Industries for 1921. Bureau of Fisheries Document No. 932. Washington, DC.

Rathbun, R. 1887. The crab, lobster, crayfish, rock lobster, and prawn fisheries. Pages 629–648 in Fisheries and Fishery Industries of the United States. Section 5, Volume 2. History and Methods of the Fisheries. Part 31, 1(a)1. Natural History and Uses of the Blue Crab. US Government Printing Office, Washington, DC.

Rau, C. 1884. Prehistoric fishing in Europe and North America. Smithsonian Contributions to Knowledge 509:1–342.

Rhodes, A., and L. Shabman. 1994. Virginia's blue crab pot fishery: The issues and the concerns. Virginia Sea Grant Report 94-09. Charlottesville, VA. 85 pages.

Richards, R.A., and P.J. Rago. 1999. A case history of effective fishery management: Chesapeake Bay striped bass. North American Journal of Fisheries Management 19:356–375.

Richmond Times-Dispatch. 1904. Charles City bacon plentiful and everybody is happy. June 5.

Ringwalt, J.L. 1888. Development of Transportation Systems in the United States. Self-published, Philadelphia. 398 pages.

Roberts, C.M. 2007. The Unnatural History of the Sea. Island Press, Washington, DC. 456 pages.

Roberts, W.A. 1905. The crab industry of Maryland. Pages 417–432 in Report of the Bureau of Fisheries 1904. Washington, DC.

Royalle, A.N. 1826. Sketches of History, Life, and Manners, in the United States. Printed for the Author, New Haven, CT. Reprinted in 1970 by Johnson Reprint Company, New York.

Rubin, J. 1961. Canal or railroad? Imitation and innovation in response to the Erie Canal in Philadelphia, Baltimore, and Boston. Transactions of the American Philosophical Society (New Series) 51(7):1–106.

S.H. [sic]. 1833. Duck shooting on the Chesapeake Bay. American Turf Register and Sporting Magazine 4(12):629–637.

Sabine, L. 1853. The principle fisheries of the American Seas. Robert Armstrong, Washington, DC. 317 pages.

Scharf, J.T. 1879. History of Maryland from the Earliest Period to the Present Day. Volume II. Hohn B. Piet, Baltimore, MD. 635 pages.

Schindler, B. 2008. Rethinking Middle Woodland settlement and subsistence patterns in the middle and lower Delaware Valley. North American Archaeologist 29:1–12.

Schulte, D.M. 2017. History of the Virginia oyster fishery, Chesapeake Bay, USA. Frontiers in Marine Science 4(127):1–19.

Schultz, F.W. 1908. Solder: Its Production and Application with a Brief History of Tin and Lead. Macneal Printing Company, Baltimore, MD. 122 pages.

Secor, D.H. 2002. Atlantic sturgeon fisheries and stock abundances during the late nineteenth century. American Fisheries Society Symposium 28:89–98.

Sette, O.E., and R.H. Fiedler. 1925. A survey of the condition of the crab fisheries of Chesapeake Bay. US Bureau of Fisheries Special Memorandum 1607-14. 19 pages, 17 tables, and 15 figures.

Sharpless, J.J. 1830. Chesapeake duck shooting. Pages 41–46 in The Cabinet of Natural History and American Rural Sports with Illustrations, Volume 1. J. & T. Doughty, Philadelphia. Reprinted in Porter (1846).

Shell Fish Commission. 1916. Seventh Report of the Shell Fish Commission of Maryland 1914 and 1915. Baltimore, MD. 78 pages.

Smith, H.M. 1891a. Notes on the crab fishery of Crisfield, MD. Bulletin US Fish Commission 9(1889):103–112.

———. 1891b. Notes on an improved form of oyster tongs. Bulletin US Fish Commission 9(1889):161–165.

———. 1893. Report on the inquiry regarding the methods and statistics of the fisheries. Pages 174–204 in Report of the Commissioner for 1889 to 1891. US Commission of Fish and Fisheries, Part 17. Washington, DC.

———. 1894. Statistics on the fisheries of the United States. Bulletin US Fish Commission for 1893, 13:389–417.

———. 1895a. Report of the division of statistics and methods of the fisheries. United States Commission of Fish and Fisheries Report of the Commissioner for the Year ending June 30, 1893, 19:52–77.

———. 1895b. A statistical report on the fisheries of the Middle Atlantic States. Bulletin US Fish Commission for 1894, 14:341–467.

———. 1914a. Shad and herring fisheries of Chesapeake Bay. Pages 65–66 in Report of the US Commissioner of Fisheries for the Fiscal Year 1913 with Appendixes. Washington, DC.

———. 1914b. Passing of the sturgeon. Pages 66–67 in Report of the US Commissioner of Fisheries for the Fiscal Year 1913 with Appendixes. Washington, DC.

———. 1917. Shad and alewife industry of Chesapeake Bay and tributaries. Pages 65–72 in Report of the US Commissioner of Fisheries for the Fiscal Year 1916 with Appendixes. Washington, DC.

Smith, J. 1608. A true relation of such occurrences and accidents of note as hath hapend in Virginia since the first planting of that collony, which is now resident in the south part thereof, till the last returne from thence. Page 43, Volume I in P.L. Barbour (ed.). 1986. The Complete Works of Captain John Smith (1580–1631) in Three Volumes. University of North Carolina Press, Chapel Hill, NC.

———. 1612. A map of Virginia with a description of the countrey, the commodities, people, government and religion. Page 155, Volume I in P.L. Barbour (ed.). 1986. The Complete

Works of Captain John Smith (1580–1631) in Three Volumes. University of North Carolina Press, Chapel Hill, NC.

———. 1624. The generall historie of Virginia, New England & the Summer Isles. Pages 103 (The Second Book) and 213 (The Third Book), Volume II in P. L. Barbour (ed.). 1986. The Complete Works of Captain John Smith (1580–1631) in Three Volumes. University of North Carolina Press, Chapel Hill, NC.

Smith, T.I.J. 1985. The fishery, biology, and management of Atlantic sturgeon, *Acipenser oxyrhynchus*, in North America. Environmental Biology of Fishes 14:61–72.

Snyder, J.P. 1917. Report of fisheries and fish cultural conditions on the Eastern Shore of Maryland. US Bureau of Fisheries Report to the Conservation Commission of Maryland. Thomas & Evans Printing Co., Baltimore, MD. 31 pages.

Starkey, D.J., P. Holm, and M. Barnard (eds.). 2008. Oceans Past: Management Insights from the History of Marine Animal Populations. Earthscan, Sterling, VA. 223 pages.

Stevenson, C.H. 1894. The oyster industry of Maryland. Bulletin US Fish Commission for 1892 12:205–297 + 15 plates.

———. 1898. The restricted inland range of shad due to artificial obstructions and its effect on natural reproduction. Bulletin US Fish Commission for 1897 17:265–271.

———. 1899a. The shad fisheries of the Atlantic coast of the United States. Report of the Commissioner for the Year Ending June 30, 1898. Part 24:101–269.

———. 1899b. The preservation of fishery products for food. Bulletin of the US Fish Commission for 1898 18:335–563.

———. 1909. Fisheries in the ante-bellum south. Pages 267–271 in J.C. Ballagh (ed.). IV The Economic History, 1607–1865. The South in the Building of the Nation, Volume V. The Southern Historical Publication Society, Richmond, VA.

Stover, J.F. 1987. History of the Baltimore and Ohio Railroad. Purdue University Press, West Lafayette, IN. 419 pages.

Strachey, W. 1610. The Historie of Travaile into Virginia Britannia. Edited in 1899 by R.H. Major. Hakluyt Society, London.

Sullivan, C.J. 2003. Waterfowling on the Chesapeake 1819–1936. Johns Hopkins University Press, Baltimore, MD. 195 pages.

Sutherland, D.R. 1974. Excavations at the Spanish Mount shell midden Edisto Island, South Carolina. South Carolina Antiquities 6(1):185–195.

Sweet, G. 1941. Oyster conservation in Connecticut: Past and present. Geographical Review 31(4):591–608.

Tarnowski, M. 2000. Chronology of factors affecting blue crab harvests. Waterman's Gazette 27(4):11, 22, 24; 27(5):11, 27; 27(6):11; 27(7):11.

Thom, W.H. de C.W. 1898. The diamond-back terrapin. Forest and Stream 50(8):151.

Thorbjarnarson, J., C. Lagueux, D. Bolze, M.W. Klemens, and A.B. Meylan. 2000. Human use of turtles: a worldwide perspective. Pages 33–84 in M.W. Klemens (ed.). Turtle Conservation. Smithsonian Institution Press, Washington, DC.

Tilp, F. 1978. This Was Potomac River. Self-published, Alexandria, VA. 358 pages.

Timmons, W.E. 1874. Report of the Commander of the Oyster Fisheries and Water Fowl of Maryland to His Excellency, the Governor, and the Commissioners of the State O. F. Force, January 1st, 1874. Wm. T. Iglehart & Co., Annapolis, MD. 10 pages.

Tinkcom, M.B. 1951. Caviar along the Potomac: Sir Augustus John Foster's "Notes on the United States," 1804–1812. The William and Mary Quarterly, Third Series 8(1):68–107.

Tower, W.S. 1908. The passing of the sturgeon: A case of the unparalleled extermination of a species. Popular Science Monthly 73:361–371.

Towle, G.M. 1870. American Society, Volume 1. Chapman and Hall, London. 336 pages.

Townsend, C.H. 1901. Statistics of the fisheries of the Middle Atlantic States. US Commission of Fish and Fisheries. Report of the Commissioner for the Year Ending June 30, 1900, 26:195–310.

True, F.W. 1887. The turtle and terrapin fisheries. 2. The terrapin fishery. Pages 499–503 in G.B. Goode (ed.). The Fisheries and Fishery Industries of the United States. Section V. History and Methods of the Fisheries, Volume 2, Part 9. US Government Printing Office, Washington, DC.

Truitt, R.V. 1939. Diamondback terrapin. Pages 93–96 in R.V. Truitt. Our Water Resources and Their Conservation. Contribution Number 27 from the Chesapeake Biological Laboratory, Solomons, MD.

Vallandigham, E.N. 1894. He raises terrapin. The Jamestown [North Dakota] Weekly Alert. November 29.

Van Engel, W.A. 1962. The blue crab and its fishery in Chesapeake Bay. Part 2: Types of gear for hard crab fishing. Commercial Fisheries Review 24(9):1–10.

———. 1998. Laws, regulations, and environmental factors and their potential effects on the stocks and fisheries for the blue crab, Callinectes sapidus, in the Chesapeake Bay region, 1880–1940. Virginia Institute of Marine Science SRAMSOE Number 347. 89 pages.

Van Metre, T.W. 1915. The fisheries of the Atlantic and Gulf coasts. Pages 179–201 in E.R. Johnson, T.W. Van Metre, G.G. Huebner, and D.S. Hanchett (eds.). History of Domestic and Foreign Commerce of the United States, Volume 2, Chapter 33. Publication 215A, Carnegie Institution of Washington, Washington, DC.

Velema, G.J., and H. Speir. 2007. The 2006 diamondback terrapin fishery in Maryland's Chesapeake Bay. Maryland Department of Natural Resources Fisheries Technical Memorandum 35:1–11.

Vincent, C.L., and J.W. Downey. 1903. Report of Clarence L. Vincent, of Worcester County, and Dr. Jesse W. Downey, of Frederick County, the Commissioners of Fisheries of Maryland for 1902–1903. Baltimore, MD. 43 pages.

W.S. [sic]. 1898. Hornaday on the destruction of our birds and mammals. Auk 15(3):280–281.

Waldman, J. 2013. Running Silver: Restoring Atlantic Rivers and Their Great Fish Migrations. Lyons Press, Guilford, CT. 304 pages.

Wallace, D.H. 1952. A critique of present biological research on oysters. Proceedings of the Gulf and Caribbean Fisheries Institute 5:132–136.

Walter, J.F., III, and J.E. Olney. 2003. Feeding behavior of American shad during spawning migration in the York River, Virginia. Pages 201–209 in K.E. Limburg and J.R. Waldman (eds.). Biodiversity, Status, and Conservation of the World's Shads. American Fisheries Society Symposium 3. Bethesda, MD.

Walter, R.C., and D.J. Merritts. 2008. Natural streams and the legacy of water-powered mills. Science 319:299–304.

Ward, D.R. 1990. Processing crustaceans. Pages 174–181 in R.E. Martin and G.J. Flick. The Seafood Industry. Van Nostrand Reinhold, New York.

Warner, W.W. 1976. Beautiful Swimmers: Watermen, Crabs and the Chesapeake Bay. Atlantic-Little, Brown, Boston. 304 pages.

Washington Post. 1880. A talk about terrapins. How Washington's favorite delicacy is obtained and stewed. Reprinted in the New York Times, December 5.

Washington Star. 1884. Terrapin and terrapin-eaters. What a Washington caterer had to tell about them. Reprinted in the New York Times, May 4.

Watts, B. D. 2013. Waterbirds of the Chesapeake: A Monitoring Plan. Version 1.0. Virginia Department of Game and Inland Fisheries, Richmond, VA. 95 pages.

Wennersten, J.R. 2007. The Oyster Wars of Chesapeake Bay, 2nd Edition. Eastern Branch Press, Washington, DC. 164 pages.

Wharton, J. 1954. The Chesapeake Bay crab industry. US Fish and Wildlife Service Fishery Leaflet 358 (Revised):1–17.

———. 1957. The Bounty of the Chesapeake: Fishing in Colonial Virginia. University Press of Virginia, Charlottesville. 78 pages.

Whitaker, A. 1624. Good news from Virginia, sent from James his towne this present moneth of March, 1623. Pages 579–588 in A. Brown. 1890. The Genesis of the United States, Volume II. The Riverside Press, Cambridge, MA.

White, A. 1633. An account of the colony of the Lord Baron of Baltamore [sic]. Available in C.C. Hall (ed.). 1910. Narratives of Early Maryland 1633–1684. Barnes and Noble, New York.

Wilberg, M.J., M.E. Livings, J.S. Barkman, B.T. Morris, and J.M. Robinson. 2011. Overfishing, disease, habitat loss, and potential extirpation of oysters in upper Chesapeake Bay. Marine Ecology Progress Series 436:131–144.

Wilkinson, J.B. 1840. The Annals of Binghamton. Cooke & Davis Printers, Binghamton, NY.

Wilson, W.T. 1977. History of Crisfield and Surrounding Areas on Maryland's Eastern Shore. Gateway Press, Baltimore, MD. 405 pages.

Winslow, F. 1881. Deterioration of American oyster-beds. Popular Science Monthly 20:29–43, 145–156.

———. 1882. Report on the oyster beds of the James River, Virginia, and of Tangier and Pocomoke Sounds, Maryland and Virginia. US Coast and Geodetic Survey for 1881, Appendix 11. Washington, DC. 87 pages.

———. 1884. Present condition and future prospects of the oyster industry. Transactions of the American Fisheries Society 13:148–163.

Wise, J.C. 1911. Ye Kingdome of Accawmacke or the Eastern Shore of Virginia in the Seventeenth Century. The Bell Book and Stationery Co., Richmond, VA. 406 pages.

Wolfe, D.A. 2000. A History of the Federal Biological Laboratory at Beaufort, North Carolina 1899–1999. National Oceanic and Atmospheric Administration, Beaufort, NC. 312 pages.

Woodhead, N. 2007. Brains and Brawn . . . Trotters and Tripe: Forgotten & Forbidden Foods from Old Cook Books. Lulu Press, Morrisville, NC. 212 pages.

Wright, A.H. 1910. Some early records of the passenger pigeon. Auk 27:428–433.

Wright, H. 1884. On the early shad fisheries of the North Branch of the Susquehanna River. Pages 619–642 in United States Commission of Fish and Fisheries, Report of the Commissioner for 1881, Part 9. Washington, DC.

Wyman, W. 1884. Hardships of the coasting trade, and particularly of the Chesapeake Bay oystermen. Public Health Papers and Reports 10:273–281.

Yates, C.C. 1913. Summary of survey of oyster bars of Maryland 1906–1912. US Coast and Geodetic Survey, Washington, DC. 81 pages.

Index